Student Study Guide in Genetics

Ken Zwicker
Whitehead Institute

Wm. C. Brown Publishers

Contents

Acknowledgments

This study guide is the conglomeration of the efforts of many people. Contributions to the content of the guide came from Bob Tamarin of Boston University and Marita Sheridan of Rhode Island College. The layout and format of this study guide is due to the desktop publishing expertise of David and Melanie Zwicker, providing camera-ready copy to Wm. C. Brown Publishers. Marge Manders of Wm. C. Brown provided the publication guidelines in which to work, as well as her understanding and patience during its completion.

Ken Zwicker

GENETICS TEXTBOOK CORRELATION CHART

This study guide follows the chapter outline of Tamarin's <u>Principles of Genetics</u>; however, it can be used in conjunction with any of the other leading introductory genetics textbooks.

CHAPTER

Zwicker, <u>Student Study Guide for Genetics,</u> 1/e, WCB, 1991 & Tamarin, <u>Principles of Genetics,</u> 3/e, WCB, 1991	Russell, <u>Genetics,</u> 2/e, Scott, Foresman/Little, Brown, 1990	Suzuki, *et al.* <u>An Introduction to Genetic Analysis,</u> 4/e, Freeman, 1989	Weaver and Hedrick, <u>Genetics,</u> 1/e, WCB, 1989	Hartl, *et al.* <u>Basic Genetics,</u> 1/e, Jones and Bartlett, 1988	Klug and Cummings, <u>Concepts of Genetics,</u> 2/e, Merrill, 1986
2. Mendel's Principles	2, 4	2, 4 12	2, 3	1	3, 4
3. Mitosis and Meiosis	1	3	4	2	2
4. Probability and Statistics	2	5	2	1, 2	3
5. Sex Determination, Sex Linkage, and Pedigree Analysis	3	3	2, 3 4	1, 2	4, 5
6. Linkage and Mapping in Eukaryotes	5, 6	5, 6	5	3	6
7. Linkage and Mapping in Prokaryotes and Viruses	7	10	6, 13	11	11, 14
8. Cytogenetics	17	8, 9	4	6	12
9. Quantitative Inheritance	23	23	3	10	7
10. Chemistry of the Gene	8, 10	11, 12	6	4, 5	8, 9 10
11. Gene Expression: Transcription	11	13	8, 9	12	16, 18
12. Gene Expression: Translation	12, 13	13	10	12	16, 17 18
13. DNA Cloning and Sequencing	14, 15	15	16	16	20, 23
14. Gene expression: Control in Prokaryotes and Viruses	18, 20	16, 19	8, 12	11, 14	21, 22
15. The Eukaryotic Chromosome	9	14	9	5	11, 19
16. Gene Expression: Control in Eukaryotes	21	16, 19 21, 22	9, 14 17	5, 14	19, 21 22
17. DNA: Its Mutation, Repair, and Recombination	16, 19	7, 17 18	7, 11	13	10, 13
18. Extrachromosomal Inheritance	22	20	15	7	15
19. Population Genetics: The Hardy-Weinberg Equilibrium and Mating systems	24	24	18	8	25
20. Population Genetics: Processes that Change Allelic Frequencies	24	24	18, 19	8, 9	25
21. Genetics of the Evolutionary Process	24	24	19	9	26

Chapter 2 - Mendel's Principles

Key Concepts and Terms

Continuous traits	Inheritance
Crossbreed	Discontinuous (discrete) traits
Self fertilization ("selfed")	Cross fertilization
Hybrid (monohybrid, dihybrid, trihybrid, ..., multihybrid)	Reciprocal cross
Filial generation (f_1, f_2, f_3,...)	Parental generation (p_1)
Recessive	Dominant
Allele	Gene
Genotype	Rule of Segregation
Homozygote	Phenotype
Progeny testing	Heterozygote
Backcross	Testcross
Partial dominance	Incomplete dominance
Wild type	Codominance
Punnett square	Mutant
Epistasis	Rule of Independent Assortment
One-gene-one-enzyme hypothesis	Hypostatic
True-breeding	Pleiotropy

Study Questions

2-1. Why was Mendel's work on inheritance not accepted by the scientific community until the turn of the century (1900's)? What caused its acceptance at that time?

Certain discoveries and theories receive what may appear to be instant acceptance (Darwin's theory of natural selection, the Crick and Watson model of DNA), while others do not (Mendel's work on the mechanisms of inheritance and Barbara McClintock's discovery of transposons). Acceptance is often dependent on how well the theory fits with currently held views of the time, and supporting evidence and mechanisms of the theory. In Mendel's case, although he had voluminous data supporting his model, there was no known cellular mechanism for his rules of inheritance. Furthermore, Mendel's mechanism of inheritance relied on discrete elements of inheritance (the gene), causing dramatic changes in the physical appearance of the individual. This seemed diametrically opposed to Darwins theory of natural selection (1959) which was published just seven years prior to Mendel's work (1966), and described change as a continuous, gradual process. Mendel's work was also very mathematical, dealing with ratios and probability, during a time when the field of biology was very descriptive and lacking in mathematical and statistical depth.

Full appreciation of Mendel's work came with the discovery of chromosomes and their behavior during cell division, which mimicked precisely Mendel's units of inheritance. Furthermore, the work of Correns, von Tschermak, and de Vries, "rediscovering" Mendel's work, provided independent verification of Mendel's principles.

A situation analogous to that of Mendel occurred in the 20th century. In the 1940's Barbara McClintock described certain genetic "autonomous elements" which move (or transpose) within the genome of corn. Like Mendel, McClintock's model of moveable "controlling elements" was without any known genetic mechanism, and her work went largely unnoticed at the time. In 1983 Barbara McClintock was awarded the Nobel Prize for her discovery of what we now call transposons (popularly referred to as "jumping genes").

2-2. List some of the characteristics that make an organism suitable for genetic study.

- ease in cultivation, handling, breeding
- short life cycle
- large numbers of progeny
- great variation in heritable traits
- ease of discriminating heritable traits

Past and present examples include pea plants, fruit flies (*Drosophila*), bacteria, fungi.

2-3. Discuss the use and features of the Punnett square.

The Punnett square is a matrix used to diagram the Mendelian genetics of a mating of two individuals. The rows of the matrix represent the possible gametic contribution of

one parent and the columns represent the possible gametic contribution of the other parent. Each cell that makes up the body of the matrix, represents the genotype of an individual due to the fusion of two particular gametes.

Implicit in the punnett square is the equal probability of the formation of each parental gamete and equal probability of the successful fusion of any two parental gametes.

Aa x Aa

	A $(p=.5)$	a $(p=.5)$
A $(p=.5)$	AA $(p=.25)$	Aa $(p=.25)$
a $(p=.5)$	Aa $(p=.25)$	aa $(p=.25)$

Possible Parental Gametes (equal probability)

genotype	probability of occurrance
AA	25%
Aa	50%
aa	25%

2-4. In an effort to develop a true-breeding strain of tall pea plants, a genetics student crossed two tall plants. All of the 200 progeny (f_1) of this cross were tall. She then chose two plants from the f_1 generation and crossed them. All of the f_2 generation were tall. She crossed two of the tall f_2 plants and produced an f_3 generation of all tall plants. However in crossing two plants of the f_3 generation she got dwarf plants in the 4th generation. She would have dismissed it as a new mutation, however a fair proportion of the f_4 plants were dwarf. Explain what happened and predict what proportion of the f_4 generation were dwarf.

This scenario illustrates the hidden nature of recessive alleles. Apparently one of the p_1 plants was heterozygous for the plant height gene, producing an f_1 generation that was phenotypically all tall but genotypically half homozygous dominant and half heterozygous. By the luck of the draw she chose a homozygous dominant plant and a heterozygote from the f_1 for crossing (a 50-50 chance).

This produced an f_2 generation of phenotypically tall plants but genotypically homozygous dominant and heterozygous. By chance again she chose a homozygote and a heterozygote for breeding, producing the same results in the f_3 generation.

Her last selection for breeding from the f_3 generation happened to be two heterozygotes (a 25% probability of occurring). This produced an f4 generation of tall and dwarf plants, with 1/4 of the plants being dwarf.

Generation	Cross for next generation	genotype
P₁	*TT x Tt*	1/2 *TT* : 1/2 *Tt*
F₁	*TT x Tt*	1/2 *TT* : 1/2 *Tt*
F₂	*TT x Tt*	1/2 *TT* : 1/2 *Tt*
F₃	*Tt x Tt*	1/4 *TT* : 1/2 *Tt* : 1/4 *tt* (dwarf plants)

Probability of Parent selection

genotype of first parent	genotype of second parent	
TT	TT	25%
TT	Tt	50%
Tt	TT	
Tt	Tt	25%

2-5. In any breeding program (cattle, horses, dogs, etc.), which is more easily eliminated from a population; a trait express by a completely dominant allele, or a trait expressed by a completely recessive allele?

Dominant alleles are much more easily eliminated in a breeding program. Since the dominant allele is always expressed phenotypically, individuals showing that phenotype can be eliminated from the breeding stock thus eliminating the allele. This results in the breeding of homozygous recessive individuals only.

2-6. Illustrate a trihybrid cross using a Punnett square. Give the phenotypic ratio, assuming complete dominance in all of the genes.

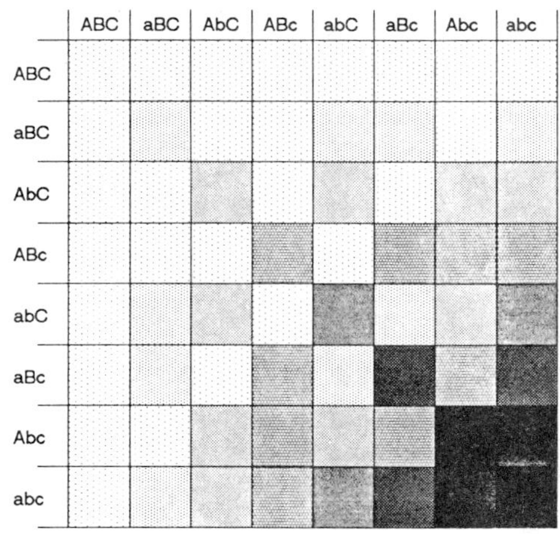

2-7. Given that humans have 46 chromosomes (23 pairs), and assuming that no other mechanism of genetic variation exists (no recombination events, no mutation), how many different types of gametes can be produced? Which of Mendel's rules determines this number? How many different zygotes are possible from a mating of two humans?

Mendel's rule of independent assortment assures us that each of the 23 pairs of chromosomes will separate independent of one another. The number of possible gametes is therefore 2^{23} or 8,388,608 and the number of possible zygotes is $(2^{23})^2$ or 7.0369×10^{13}.

2-8. Complete the following table of all of the possible blood types of children from each of the following matings of parental blood types.

		MOTHER'S BLOOD TYPE			
		A	B	AB	O
FATHER'S BLOOD TYPE	A				
	B				
	AB				
	O				

First consider the possible genotypes that would produce a given phenotype, then consider all of the possible combinations of those alleles.

MOTHER'S BLOOD TYPE

Possible Genotype	A (I^AI^A)	A (I^AI^O)	B (I^BI^B)	B (I^BI^O)	AB (I^AI^B)	O (I^OI^O)
A I^AI^A	A	A	AB	A / AB	A, AB	A
A I^AI^O	A	A / O	B / AB	A, B / AB, O	A, B, AB	A, O
B I^BI^B			B	B	B, AB	B
B I^BI^O			B	B / O	A, B, AB	B, O
AB I^AI^B					A, B, AB	A, B
O I^OI^O						O

(FATHER'S BLOOD TYPE)

2-9. How many genotypes are possible (in a diploid organism) from a single gene if there exists;

a) 2 alleles for the gene?

b) 3 alleles for the gene?

c) 4 alleles for the gene?

```
a)  3            AA, Aa,
                    aa
b)  6         AA, Aa1, Aa2,
            a1a1, a1a2,
                  a2a2
c)  10     AA, Aa1, Aa2, Aa3,
          a1a1, a1a2, a1a3,
              a2a2, a2a3
                a3a3
```

2-10. Pea pod shape is controlled by a one gene - two allele genetic system. Full pod shape is completely dominant to constricted pea pod shape. Pea plant size is also controlled by a one gene - two allele system with tall size being completely dominant to dwarf size. Given a plant that was heterozygous for pod shape and homozygous recessive for plant size:

a) What is the phenotype of this plant?

b) If the this plant were crossed with a dihybrid , what would be the phenotypic ratio of pod shape?

c) What would be the phenotypic ratio of plant size?

d) What would be the phenotypic ratio of pod shape and plant size together?

a) The plant would be a full pod dwarf plant (*Fftt*)

b) The cross would be *Fftt* x *FfTt*, however since this question only deals with pod shape, we need only concern ourselves with that part of the genotype

Ff x Ff

	F	f
F	FF	Ff
f	Ff	ff

the phenotypic ratio is 3 full : 1 constricted

c) Here we need only concern ourselves with the plant size genotype

tt x Tt

	T	t
t	Tt	tt

the phenotypic ratio is 1 tall : 1 dwarf

d) Now we must take into consideration all possible combinations of both traits together

Fftt x FfTt

	FT	Ft	fT	ft
Ft	FFTt	FFtt	FfTt	Fftt
ft	FfTt	Fftt	ffTt	fftt

the phenotypic ratio is 3 full pod tall : 3 full pod dwarf : 1 constricted pod tall : 1 constricted pod dwarf.

2-11. Given the following mating

```
p1                      AA    x    aa
                      (male)    (female)


f1                          Aa
                         (female)
```

illustrate the following crosses;

a) reciprocal cross homozyous recessive X homozygous dominant

b) back cross – homozygous recessive/dominant X heterozygous

c) test cross – homozygous recessive X heterozygous

```
  a)   p1                aa    x    AA
                       (male)    (female)
  b)   p1                AA    x    Aa      or      aa    x    Aa
                       (male)    (female)         (male)    (female)
  c)   p1                aa    x    Aa
                       (male)    (female)
```

2-12. One form of congenital blindness (retinitis pigmentosa) is controlled by 2 genes each with two alleles. Both genes exhibit complete dominance, but one is epistatic to the other. In the epistatic gene the mutant allele (that which would cause blindness) is dominant to the wild type (normal sighted) allele. In the hypostatic gene the mutant allele is recessive to the wild type.

a) What would be the phenotype of a woman heterozygous for both genes (sighted or blind)?

b) What would be the genotype of a blind man whose parents were both sighted?

c) What would be the genotypes of his parents?

d) If the heterozygous woman of part a married the blind man of part b and wanted to know if they could have any normally sighted children, what would you tell them?

The first step here is to set up the genetic system given the information provided. We will use the following designations:

E = dominant mutant allele of the epistatic gene

e = recessive wild type allele of the epistatic gene

H = dominant wild type allele of the hypostatic gene
h = recessive mutant allele of the hypostatic gene

The relationship of genotype to phenotype for this 2 gene, 2 allele system would be the following;

Genotype	Phenotype
$E\_H\_$	blind
$eeH\_$	sighted
E_hh	blind
$eehh$	blind

a) A woman heterozygous for both genes (*EeHh*) would be afflicted with retinitis pigmentosa since the mutant allele is dominant for the epistatic gene.

b) Both sighted parents must have the genotype *eeH_*, therefore their son must be homozygous recessive for the epistatic gene (since each parent is only capable of contributing the recessive allele). Knowing this, the only way the son could be afflicted is if he were homozygous for the hypostatic gene. His genotype must be *eehh*.

c) Since the son is homozygous recessive for the hypostatic gene, each parent must have contributed a recessive allele

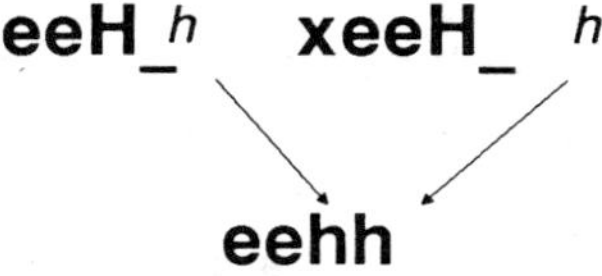

d) the cross would look like this;

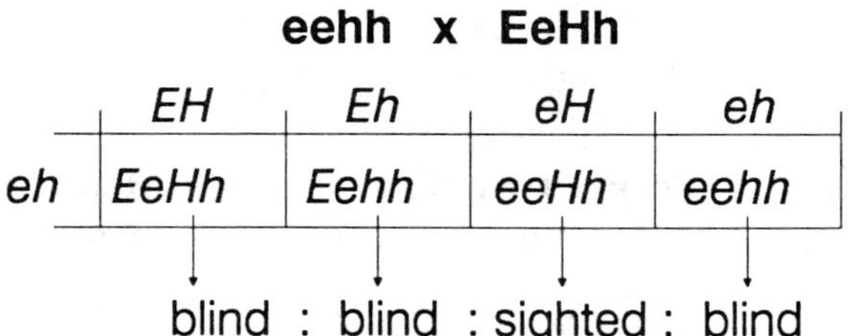

the couple would have a 1 in 4 chance of having a normally sighted child.

2-13. Provide reasonable allelic designations for these "unreasonable" fruit fly traits.

Propeller mutant allele which is recessive
Dominant wild type allele of a long-legged mutation
Dominant mutant allele of a feathered mutation
Wild type allele of a recessive "killer" mutation
Recessive wild type allele of a tail mutation
Wild type allele of a dominant toothed mutation
Mutant allele of a recessive bipedal mutation
Mutant allele of a dominant chlorophyll mutation

	Allelic designation
Propeller mutant allele which is recessive	$prop$
Dominant wild type allele of a wheeled mutation	$lleg^+$
Dominant mutant allele of a feathered mutation	Fea
Wild type allele of a recessive "killer" mutation	klr^+
Recessive wild type allele of a tail mutation	Tl^+
Wild type allele of a dominant toothed mutation	Tth^+
Mutant allele of a recessive bipedal mutation	bp
Mutant allele of a dominant chlorophyll mutation	Chl

2-14. The following is the biochemical pathway for pyrimidine biosynthesis.

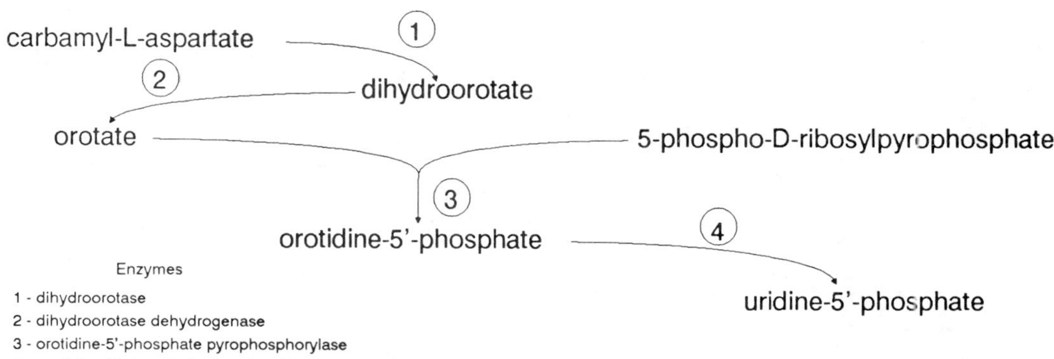

The following *E. coli* **mutants were grown in a series of culture plates containing growth media with various additives.**

mutant 1 - defective orotidine-5'-phosphate decarboxylase
mutant 2 - wild type (it is not really a mutant at all)
mutant 3 - cannot produce 5-phospho-D-ribosylpyrophosphate
mutant 4 - defective dihydroorotate dehydrogenase
mutant 5 - defective orotidine-5'-phosphate pyrophosphorylase
mutant 6 - defective dihydroorotase

Using + to indicate growth, and - to indicate no growth, complete the following table predicting what results you would expect.

additive

mutant	minimal media	dihydroorotate	orotate	orotidine-5'-phosphate	uridine-5'-phosphate
1					
2					
3					
4					
5					
6					

additive

mutant	minimal media	dihydroorotate	orotate	orotidine-5'-phosphate	uridine-5'-phosphate
1	-	-	-	-	+
2	+	+	+	+	+
3	-	-	-	+	+
4	-	-	+	+	+
5	-	-	-	+	+
6	-	+	+	+	+

2-15. Radio-astronomers have just received signals from a galaxy far far away (and obviously a long time ago) from a species of beings (the "trinoids") in a trinary solar system (containing three suns). The first information this culture decided to broadcast was information concerning the biology of their planet. Surprisingly, their biological system is remarkably similar to ours with some very interesting differences in their genetics. The species requires three sexes for reproduction (male, female, and hemale) and is triploid (three copies of each gene instead of two), with 23 triplets of chromosomes (69 chromosomes) instead of 23 pairs as we have. Other than that, the history of their planet parallels ours exactly.

a) They too had a Mendel on their planet. What would his rules of segregation and independent assortment be?

b) Their Reginald Punnett also devised a matrix to calculate possible genotypic outcomes from matings. How is it different from ours (if at all)?

c) Can you come up with a general formula for determining the number of possible gametes produced by a hybrid organism, and the proportion of homozygous recessives produced by a hybrid cross in a three allele system?

a) The trinoids rule of segregation would be: "A gamete receives only one allele from the triplet of alleles possessed by an individual; fertilization (the union of three gametes) reestablishes the triple number." The trinoids rule of independent assortment would be the same as ours: "Alleles of one gene segregate independently of alleles of other genes."

b) The trinoid punnett square would actually be a "punnett cube." Each dimension of the cube would be defined by the possible gametes of each of the three sexes

c) The constant 2 in our general expressions 2^n and $1/(2^n)^2$ is due to the fact that we are dealing with a two-sex, diploid genetic system. For the trinoids, the constant would be 3. The general formula for the number of possible gametes of a hybrid would be 3^n. the formula for the proportion of homozygous recessives of a hybrid cross would be $1/(3^n)^3$

$$Aa_1a_2 \times Aa_1a_2 \times Aa_1a_2$$

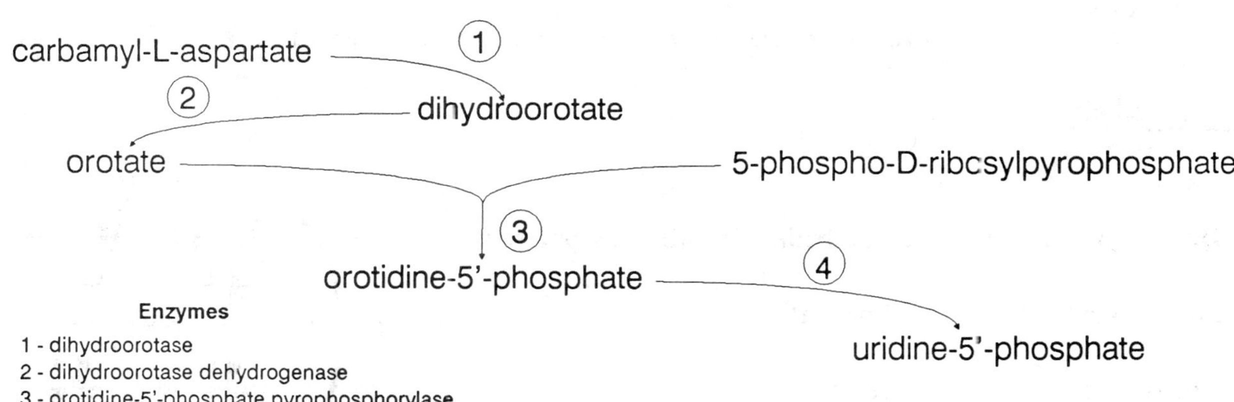

An individual with the genotype AAa_1, Aa_1a_1, Aa_2a_2 would not be considered true heterozygotes, but may be referred to as either "heterozygous dominant" or "heterozygous recessive" respectively.

Key Concepts and Terms

Mitosis	Karyokinesis
Cytokinesis	Meiosis
Eukaryote	Locus (loci)
Diploid ($2n$)	Prokaryote
Homologous chromosomes	Haploid (n)
idiogram	karyotype
heteromorphic chromosome	homomorphic chromosome
Prophase (I, II)	Interphase
Anaphase (I, II)	Metaphase (I, II)
Cell cycle (S, G_1, M, G_2)	Telophase (I, II)
Centrioles	Nucleolus, nucleolar organizer
Chromatid	Spindles, microtubules
Metaphase plate	Sister chromatids
Synaptonemal complex	Synapsis
Tetrad	Bivalent
Dyad	Chiasma
Recombination	Crossing over
Equational division	Reductional division
Spermatogonium	Spermatogenesis
Oogonia	Oogenesis
Primary spermatocyte	Primary oocyte
Secondary oocyte	Secondary spermatocyte
Spermatid	Polar body
Ovum	Sperm
Sporophyte	Parthenogenesis
The chromosomal theory of inheritance	Gametophyte
Gene mapping	Linkage

Chromosome (chromatin, kinetochore, centromere, metacentric, telocentric, acrocentric)

Study Questions

3-1. Draw a chromosomal figure, labeling the following parts.

chromosome	**bivalent**
centromere	**tetrad**
chromatid	**dyad**
sister chromotids	**homologous chromosomes**

Give the diploid number ($2n$) and haploid number (n) of your figure.

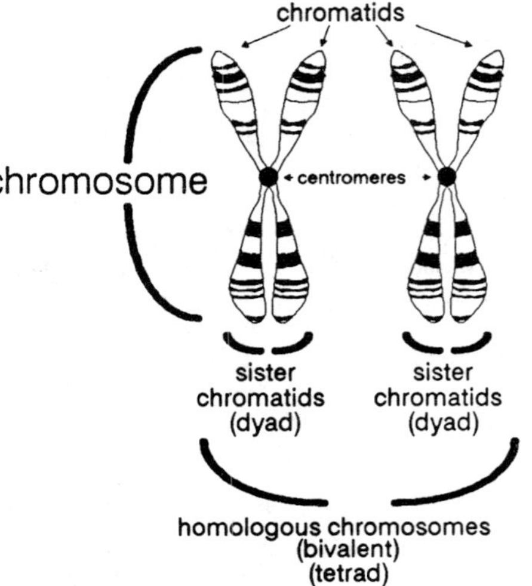

Given the figure below of the primary spermatocyte and primary oocyte, answer questions 2 through 6.

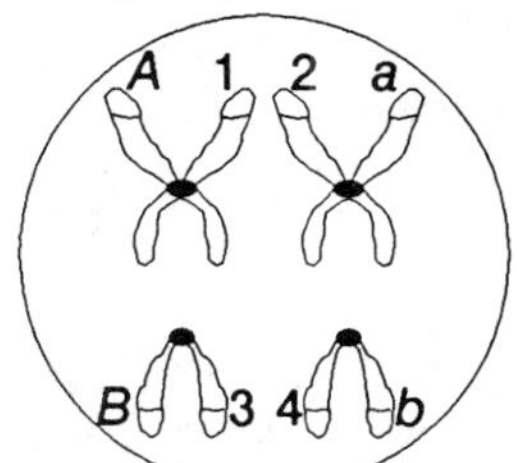

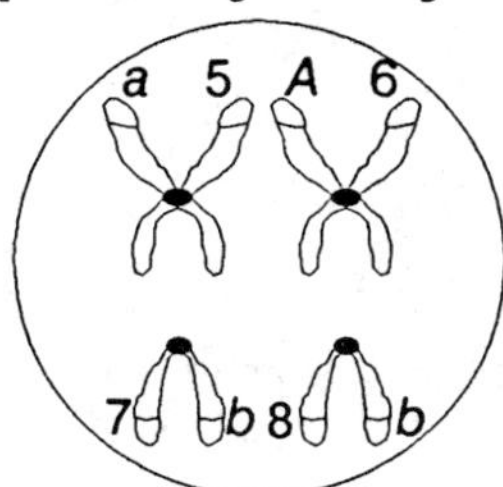

3-2. What stage of meiosis are these cells at, and what is the diploid number of this species?

The cells are in metaphase I of meiosis and have a $2n$ of 4.

3-3. The "A" and "B" gene locations (loci) are identified on the chromosomes with some information missing. Identify the specific allele (A or a; B or b) at sites 1 through 8.

The allele on one chromatid is identical to that of its sister chromatid (barring mutation).

1.	*A*	5.	*a*
2.	*a*	6.	*A*
3.	*B*	7.	*b*
4.	*b*	8.	*b*

3-4. What is the genotype of the male that produced the spermatocyte and the female that produced the oocyte?

The genotype of the male is *AaBb* and the female is *Aabb*.

3-5. Finish drawing all the possible cellular outcomes of the meiotic division for the spermatocyte and the oocyte.

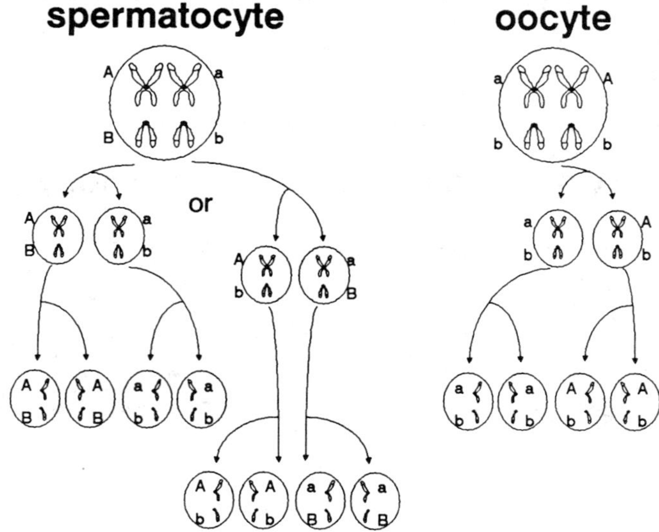

3-6. During which division of meiosis do Mendel's rules of segregation and independent assortment apply (use the figure you made for question 3-5 as a guide)?

Genes (located on different chromosomes) segregate <u>and</u> assort independently during the first (reductional) division of meiosis. This is represented in the diagram of question 3-5 in which there are two possible pathways the primary spermatocyte can take. Either the *A* and *B* alleles co-migrate (and, therefore, also the *a* and *b* alleles) during karyokinesis, or the *A* and *b* alleles co-migrate (and, therefore, the *a* and *B* alleles).

Since the female oocyte is homozygous for the *b* allele, it is irrelevant which direction the chromosome of the "*B*" gene migrates.

3-7. Identify the number of chromosomes and chromatids per cell by the end of each of the phases of mitosis and meiosis for an organism of *n* = 23.

<u>MITOSIS</u>	<u># OF CHROMOSOMES</u>	<u># OF CHROMATIDS</u>
Interphase		
Prophase		
Metaphase		
Anaphase		
Telophase		

MEIOSIS	# OF CHROMOSOMES	# OF CHROMATIDS
Interphase		
Prophase I		
Metaphase I		
Anaphase I		
Telophase I		
Prophase II		
Metaphase II		
Anaphase II		
Telophase II		

The number of chromosomes is defined by the number of centromeres. During anaphase of mitosis and anaphase II of meiosis the centromeres divide, which, by definition, doubles the chromosome count. There is no increase in chromosomal material though, as illustrated by the constancy in the number of chromatids. By the end of telophase the daughter cells are completely separated, halving the amount of genetic material per cell.

MITOSIS	# OF CHROMOSOMES	# OF CHROMATIDS
Interphase	46	92
Prophase	46	92
Metaphase	46	92
Anaphase	92	92
Telophase	46	46

MEIOSIS	# OF CHROMOSOMES	# OF CHROMATIDS
Interphase	46	92
Prophase I	46	92
Metaphase I	46	92
Anaphase I	46	92
Telophase I	23	46
Prophase II	23	46
Metaphase II	23	46
Anaphase II	46	46
Telophase II	23	23

3-8. The location of two genes on the same chromosome would appear to violate Mendel's rule of independent assortment. What term is used to describe this situation?

Linkage refers to the association of two or more genes to each other due to their physical proximity on a particular chromosome.

3-9. What chromosomal phenomenon lessens the co-migration of loci located on the same chromosome, and what is the significance of this phenomenon?

> **Crossing over** is the process whereby segments from homologous chromosomes are exchanged. Crossing over can result in the recombination of alleles located on the same chromosome, thus greatly increasing the number of possible gametic genotypes (genetic variability).

3-10. Given an organism with 3 chromosomes and a total of only 300 polymorphic genes (genes with two different alleles), calculate the number of different gametes that could arise. If crossing over did not occur during meiosis, how many different gametic genotypes could be produced?

> Assuming 100% independent assortment of loci, 2^{300}, or 2.037×10^{90} different gametes are possible in the meiotic process. If crossing over did not occur, genetic variability would be limited to the number of possible chromosomal rearrangements, resulting in 2^3, or 8 different genotypic gametes possible.

3-11. How many different types of zygotes are possible from the mating of two organisms from question 3-10, heterozygous for every locus assuming;

a) 100% independent assortment

b) no crossing over

> a) $(2^{300})^2$ or 2^{600}

> b) $(23)^2$ or 2^6

3-12. At metaphase I of human spermatogenesis each spermatocyte contains;

a) how many chromosomes

b) how many tetrads

c) how many chromatids

d) how many centromeres

> a)   46
> b)   22 plus the X and Y chromosomes
> c)   92
> d)   46

3-13. The common housefly (*Musca domestica*) has a haploid number of 6: three long chromosomes, one metacentric, one acrocentric, and one telocentric; three short chromosomes, one metacentric, one acrocentric, and one telocentric. Draw a typical housefly primary oocyte in metaphase I.

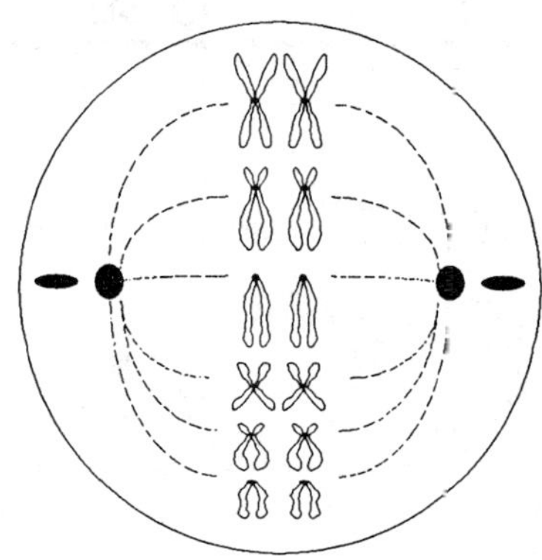

3-14. Recall our hypothetical civilization of trinoids from question 2-14. Given what you know about the process of meiosis and the orderly way in which chromosomes are segregated and assorted here on Earth, illustrate a possible mechanism whereby "trinoid meiosis" may occur in which a triploid cell gives rise to 6 haploid gametes (hint: use 3 chromosomes to show a typical "homologous triplet" - one from each parental sex)

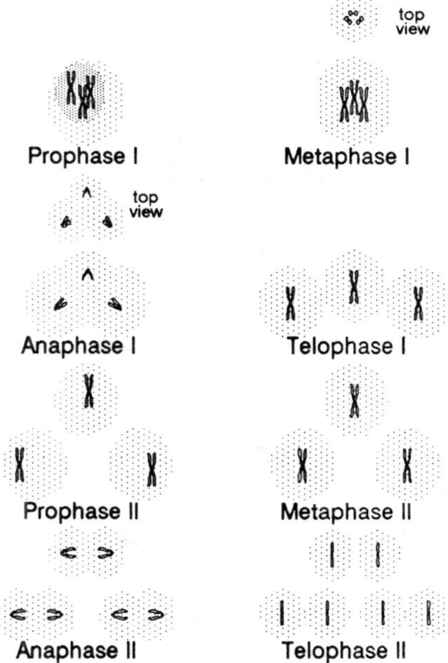

3-15. What is the chromosomal theory of inheritance?

The synthesis of the chromosomal theory of inheritance occurred in the early 1900s. The then recently revealed cellular behavior of chromosomes coincided precisely with Mendel's principles of inheritance, providing the much needed cellular mechanism for Mendel's theory of inheritance to be accepted. The theory simply states that the units of inheritance (genes) are located on chromosomes.

Key Concepts and Terms

Stochastic

Probability $(p=a/n)$

A priori probabilities

Product rule of probability ("and" rule)

Binomial expansion (multinomial expansion)

Confidence limits

Testing of hypotheses

Type I error

Chi-square distribution

Null hypothesis

Critical Chi-square value

Scientific method

Probability theory

Empirical probabilities

Sum rule of probability ("either/or" rule)

Binomial theorem $[p=(n!/s!t!)p^s q^t]$

Factorial

Experimental design

Sampling distribution

Type II error

Degrees of freedom

Level of significance

Study Questions

4-1. Assuming an equal probability of having a son or a daughter, calculate the probability of the following family cases.

a) first born being a son
b) having two sons
c) having three sons
d) having five daughters
e) a couple, who already have five daughters, having a sixth daughter
f) having a son and a daughter

a) p=0.5
b) $p=(0.5)(0.5)=0.25$
c) $p=(0.5)(0.5)(0.5)=(0.5)^3=0.125$
d) $p=(0.5)^5=0.03125$
e) $p=0.5$ This question illustrates the point that "chance has no memory." In these cases past events do not influence future probabilities
f) Since no order was specified, there are two out of four possibilities, all of equal probability.

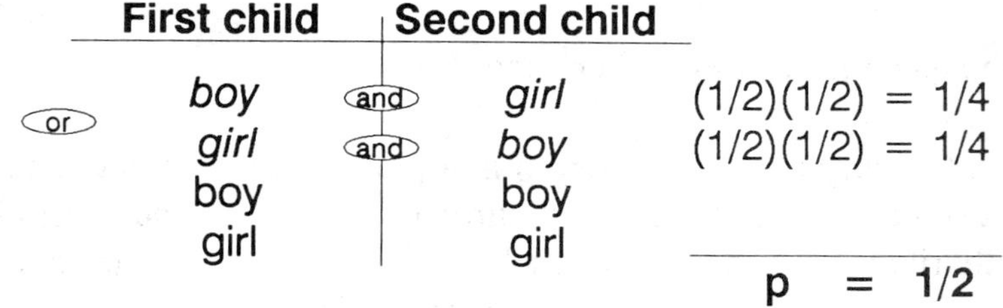

Also, since no order was specified, the binomial expansion formula can be used.

$$P=(n!/s!t!)p^s q^t$$

where;

n = total number of events (children) specified - 2
s = number of events (children) of one category (sex) - 1
t = number of events (children) of second category (sex) - 1
p = probability of the s event occurring - 0.5
q = probability of the r event occurring - 0.5

therefore;

$$P=(2!/1!1!)(0.5)^1(0.5)^1$$

$$P=(2)(0.25)$$

$$P=0.5$$

This is the simplest case use of the binomial theorem.

4-2. What is the probability that two siblings will have identical genotypes (hint: assume that no mutation or recombination occurs)?

In this problem the first child can have any genotype at all. The second child, however, must be of the same genetic make-up as the first, That is, whatever alleles were past on to the first child by the mother and father, must be passed on to the second child as well. This means that the genetic contents of the two eggs from the mother that produced the siblings must be identical, and the genetic contents of the sperm produced by the father must be identical.

Assuming that mutation and recombination is negligible, the probability is that of identical chromosomal segregation during spermatogenesis, and identical chromosomal segregation during oogenesis. Given that humans have 23 pairs of homologous chromosomes, the probability is;

$$(1/2)^{23} \quad \times \quad (1/2)^{23} \quad = \quad (1/2)^{46} = 1.42 \times 10^{-14}$$

$$\text{probability of} \quad \text{probability of} \quad = \quad \text{probability of}$$
$$\text{identical sperm} \quad \text{identical eggs} \quad \text{identical zygotes}$$

4-3. What is the probability that the eldest girl in a three- sib family will have at least one younger brother?

Since the first child is already determined to be a girl, only the sex of the two remaining siblings are of concern. There are three possible cases for the girl to have a younger brother.

Second child		**Third child**	
boy	and	*girl*	$(1/2)(1/2) = 1/4$
girl	and	*boy*	$(1/2)(1/2) = 1/4$
boy	and	*boy*	$(1/2)(1/2) = 1/4$

$$p = 3/4$$

Perhaps the quickest way of calculating this probability is to examine the single alternative case, the probability of the girl having two sisters (<u>no</u> younger brother). This probability is $(1/2)^2$ or 1/4. The remainder of the time (3/4 or 75%) the sibship will include at least one boy.

Given that albinism is a recessive trait that is epistatic to coat color in mice (in which the agouti allele is completely dominant to the black allele), and that the birth sex ratio in mice is 1:1, answer the following questions.

4-4. What is the probability that two dihybrid mice will have only one *agouti* pup that is male in a litter of four?

In this problem two factors must be considered; the probability of having a particular coat color, and the probability of the sex. Since there is no birth order with a litter, the binomial expansion is employed.

Phenotypic ratio

$$
\begin{array}{lll}
A\_C\_ & 9 & \rightarrow \text{agouti } (9) \\
a\overline{a}C\_ & 3 & \rightarrow \text{black } (3) \\
A\_c\overline{c} & 3 & \rightarrow \text{albino } (4) \\
a\overline{a}cc & 1 &
\end{array}
$$

Probability of only one agouti, male pup in a litter of four

$$p = \frac{4!}{1!\,3!} \left[\left(\frac{9}{16}\right)\left(\frac{1}{2}\right)\right]^1 \left(\frac{7}{16}\right)^3$$

probability of being agouti   probability of being male   probability of being anything but agouti (no sex specified)

The expression above reduces to $p = 0.094$.

4-5. What is the probability that two dihybrid mice will have two agouti pups, one black pup, and

$$p = \frac{4!}{2!\,1!\,1!}\left(\frac{9}{16}\right)^2\left(\frac{3}{16}\right)^1\left(\frac{4}{16}\right)^1$$

probability of agouti pup   probability of black pup   probability of albino pup

one albino in a litter of four?

The expression above reduces to $p = 0.178$.

4-6. What is the probability that two dihybrid mice will have a litter of 4 consisting of two males;

$$p = \frac{4!}{1!\,1!\,1!\,1!}\left[\left(\frac{9}{16}\right)\left(\frac{1}{2}\right)\right]^1\left[\left(\frac{3}{16}\right)\left(\frac{1}{2}\right)\right]^1\left[\left(\frac{9}{16}\right)\left(\frac{1}{2}\right)\right]^1\left[\left(\frac{4}{16}\right)\left(\frac{1}{2}\right)\right]^1$$

probability of agouti male   probability of black male   probability of agouti female   probability of albino female

one agouti, one black, and two females; one agouti, one albino?

the expression above reduces to $p = 0.0223$

Problems 4-7 through 4-15 follow a continuing line of questions. Start with 4-7 and work your way through the problems using the information provided.

As a student of genetics, you become interested in the phenotype of feathered legs found in Black Langshan fowl and featherless legs found in Buff Rocks fowl. Your first inclination is that this trait is controlled by a one gene - two allele genetic system. To test this hypothesis you mate a true breeding Black Langshan with a true breeding Buff Rocks, and all of the offspring develop feathered legs.

4-7. Outline the proposed genotypes of this mating. Do the results support your hypothesis?

```
P1                        FF          x            ff
                      (feathered)          (featherless)
F1                              Ff
                          (feathered)
```

The data are consistent with the proposed hypothesis of a one gene - two allele system.

After your first breeding experiment you become more convinced that you are dealing with a single locus system with the feathered allele (F) showing complete dominance over the featherless allele (f). Continuing the breeding study, you breed two of the f_1 birds.

4-8 What is your working hypothesis for this f_1 cross? Diagram the cross and expected results.

Your hypothesis is that the f_2 generation will consist of feathered and featherless fowl in a ratio of 3:1 respectively.

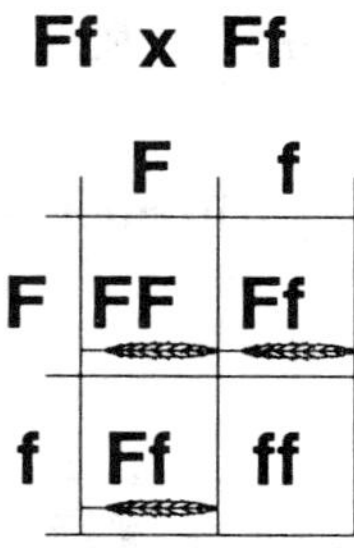

Upon performing the f_1 cross, the f_2 generation consists of 10 individuals, all of whom are feathered.

4-9. Given the proposed genetic system, what is the probability that all 10 of the f_2 generation would be feathered?

The probability of getting a feathered offspring is 0.75 and the probability of getting a featherless offspring is 0.25 (from the proposed 3:1 ratio of question 4-8). Therefore the probability of getting 10 birds, all of whom are feathered, is $(0.75)10$ or 0.056 (5.6% of the time).

You explain the lack of any featherless individuals in the f_2 generation as statistical error due to a small sample size. You increase your sample size by cross-mating several of the f_1 generation. The f_2 generation now consists of 348 eggs.

4-10. How many feathered and featherless individuals do you expect from this sample?

```
feathered fowl  - (0.75)(348) = 261
featherless fowl - (0.25)(348) =  87
                  total          348
```

After the f_2 generation matures, you find that it consists of 325 feathered individuals and 23 featherless individuals.

4-11. a) State the null hypothesis for this experiment.

b) Calculate the Chi-square value.

c) Determine the degrees of freedom for this analysis.

a) The null hypothesis is that the only difference between the observed numbers (325 feathered, 23 featherless) and the expected numbers (261 feathered, 87 featherless) is due to chance variation.

b)

Category	Observed (O)	Expected (E)	(O-E)	$(O-E)^2$	$\dfrac{(O-E)^2}{E}$
Feathered	325	261	64	4096	15.69
Featherless	23	87	-64	4096	47.08
				$x^2 =$	62.77

c) Since there are two categories (feathered or featherless), there is only one degree of freedom. In other words, for a total of 348 birds, once it is determined that 261 are feathered, the remaining 87 individuals must be featherless (or *visa versa*).

Using the Chi-square table below, answer the following questions.

Chi-square Values

Degrees of Freedom	Probabilities						
	0.99	0.95	0.80	0.50	0.20	0.05	0.01
1	0.000	0.004	0.064	0.455	1.642	3.841	6.635
2	0.020	0.103	0.446	1.386	3.219	5.991	9.210

4-12. a) What is the critical chi-square value?

b) What probability value does your chi-square value give you?

c) State completely what this probability level means in terms of your null hypothesis. Does the analysis support or refute your null hypothesis?

a) For the 0.05 level of significance, with one degree of freedom, the critical chi-square value is 3.841.

b) For a chi-square value of 62.77 and one degree of freedom, the probability level is much less than 0.01 (it is off the chart).

c) There is a much less than 0.01 probability (>1% chance) that the difference between the observed numbers and those expected is due to chance. You must therefore reject your null hypothesis. Something other than chance is probably responsible for the difference between the observed and expected numbers.

4-13. Given that out of 348 f_2 individuals, 325 were feathered and 23 were featherless, can you propose an alternative hypothesis?

The ratio of the phenotypes, 325:23 or 14:1, is close to a ratio of 16^{ths}, which is the expected ratio of a two locus, two allele system for a dihybrid cross (remember the typical 9:3:3:1 ratio). Perhaps the genetic control of feathered legs is due to two loci with one epistatic to the other giving a ratio of 15:1.

Phenotypic ratio

A_C_	9		
aaC_	3	feathered	(15)
A_cc	3		
aacc	1	featherless	(1)

4-14. Given your alternate hypothesis;

a) Calculate alternate expected values for the f_2 generation.

b) State an alternate null hypothesis.

c) Calculate an alternate chi-square value.

d) Determine the probability level of your null hypothesis.

e) Does the chi-square analysis support or refute the null hypothesis?

a) The expected values from a sample of 348 individuals, assuming a 15:1 ratio is;

$$(15/16)(348) = 326^*$$
$$(1/16)(348) = 22^*$$

$$^*\text{rounded to the nearest individual}$$

b) The difference between the observed numbers (325 feathered, 23 featherless) and those expected (326 feathered, 22 featherless) is due to chance variation.

c)

Category	Observed (O)	Expected (E)	(O-E)	$(C-E)^2$	$\frac{(O-E)^2}{E}$
Feathered	325	326	-1	1	0.003
Featherless	23	22	1	1	0.045
				$x^2 =$	0.048

d) A chi-square value of 0.048 with one degree of freedom represents a probability value less than 95% but greater than 80% ($0.95p0.8$). There is, therefore, a greater than 80% probability that the difference between the observed and expected numbers is due to chance.

e) The 0.8 probability is certainly greater than the 0.05 level of significance, therefore the chi-square analysis supports the alternate null hypothesis.

4-15. Does the preceding analysis prove your epistatic model of gene control for feathered legs?

No, in general, experimental testing and statistical analyses may <u>disprove</u> hypotheses, but they can never <u>prove</u> them. Any experimental procedure can, at best, <u>support</u> hypotheses. Also remember that in the preceding analysis, we may be committing a type II error, that is, supporting a null hypothesis that is in reality false. An alternate hypothesis has been established which now requires independent testing.

Key Concepts and Terms

Homogametic (homomorphic)
Polyploids
Nondisjunction
Genic balance theory
Doublesex (*dsx*)
Sex switch
Numerator elements
H-Y antigen (H-Y histocompatibility)
ZFY gene
Sex linkage (X-linked)
Holandric traits
Hemizygous
Crisscross pattern of inheritance
Sex-influenced (sex-controlled) traits
Penetrance
Phenocopy
Sibling (sib)
Consanguineous
Dosage compensation
Heterochromatin
Mosaicism
Allozyme

Sex chromosomes
Heterogametic (heteromorphic)
Aneuploids
Autosomes
Intersex (metamale, metafemale)
Transformer (*tra*)
Sex-lethal (*Sxl*)
Denominator elements
Testis-determining factor (*TDF*)
Od gene
Y-linked
Pseudoautosomal
Pseudodominance
Sex-limited traits
Pedigree
Expressivity
Affected individual
Propositus (proposita, proband)
Incestuous
Barr body
Lyon hypothesis
Electrophoresis
Isozyme

Study Questions

5-1. Identify the sex of the following individual organisms.

ORGANISM	# AND TYPE OF SEX CHROMOSOMES	PLOIDY OF AUTOSOMAL CHROMOSOME	SEX
Human ($2n=46$)	1 X chromosome	$2n$	
Cricket ($2n=22$)	1 X chromosome	$2n$	
Human ($2n=46$)	2 heteromorphic	$2n$	
Fruit fly ($2n=8$)	2 heteromorphic	$2n$	
Pigeon ($2n=80$)	2 heteromorphic	$2n$	
Human ($2n=46$)	2 homomorphic	$2n$	
Fruit fly ($2n=8$)	2 homomorphic	$2n$	
Cricket ($2n=22$)	2 homomorphic	$2n$	
Pigeon ($2n=80$)	2 homomorphic	$2n$	
Human ($2n=46$)	3 homomorphic	$2n$	
Fruit fly ($2n=8$)	3 homomorphic	$2n$	
Cricket ($2n=22$)	3 homomorphic	$2n$	
Human ($2n=46$)	3 X, 1 Y	$2n$	
Fruit fly ($2n=8$)	3 X, 1 Y	$2n$	
Human ($2n=46$)	2 X, 1 Y	$3n$	

Fruit fly $(2n=8)$	2 X, 1 Y	$3n$
Human $(2n=46)$	1 X, 2 Y	$2n$
Fruit fly $(2n=8)$	1 X, 2 Y	$2n$
Human $(2n=46)$	1 X, 2 Y	$3n$
Fruit fly $(2n=8)$	1 X, 2 Y	$3n$

ORGANISM	# AND TYPE OF SEX CHROMOSOMES	PLOIDY OF AUTOSOMAL CHROMOSOMES	SEX
Human $(2n=46)$	1 X chromosome	$2n$	female
Cricket $(2n=22)$	1 X chromosome	$2n$	female
Human $(2n=46)$	2 heteromorphic	$2n$	male
Fruit fly $(2n=8)$	2 heteromorphic	$2n$	male
Pigeon $(2n=80)$	2 heteromorphic	$2n$	female
Human $(2n=46)$	2 homomorphic	$2n$	female
Fruit fly $(2n=8)$	2 homomorphic	$2n$	female
Cricket $(2n=22)$	2 homomorphic	$2n$	female
Pigeon $(2n=80)$	2 homomorphic	$2n$	male
Human $(2n=46)$	3 homomorphic	$2n$	female
Fruit fly $(2n=8)$	3 homomorphic	$2n$	metafemale
Cricket $(2n=22)$	3 homomorphic	$2n$	female
Human $(2n=46)$	3 X, 1 Y	$2n$	male
Fruit fly $(2n=8)$	3 X, 1 Y	$2n$	metamale
Human $(2n=46)$	2 X, 1 Y	$3n$	male
Fruit fly $(2n=8)$	2 X, 1 Y	$3n$	intersex
Human $(2n=46)$	1 X, 2 Y	$2n$	male
Fruit fly $(2n=8)$	1 X, 2 Y	$2n$	male
Human $(2n=46)$	1 X, 2 Y	$3n$	male
Fruit fly $(2n=8)$	1 X, 2 Y	$3n$	metamale

5-2. In female kangaroos, The paternal sex chromosome is always the inactivated chromosome. Given this how would a rare sex-linked dominant trait in a male be expressed:

a) in his daughters.

b) in his sons.

c) in his daughters' daughters.

d) in his daughters' sons.

e) in his sons' daughters.

f) in his sons' sons.

The parental male would be affected with the trait, since it occurs on his one and only X chromosome. However;

a) his daughters would be phenotypically unaffected because the affected X chromosome from their father would be inactivated. These females would be carriers.

b) his sons would be unaffected and noncarriers because his contribution to their sex chromosomes would be the Y chromosome.

c) Since the parental males' daughters are carriers of the sex-linked trait, the probability is 50% that the f_2 females would receive the dominant trait from their mother, and thus be affected.

d) Similar to part c, half the time the f_1 females would contribute the affected X chromosome to their sons.

e) + f) Since the parental male contributes his Y chromosome to all of his sons, the dominant sex-linked allele is not present in his sons' offspring.

The figure below outlines the pedigree.

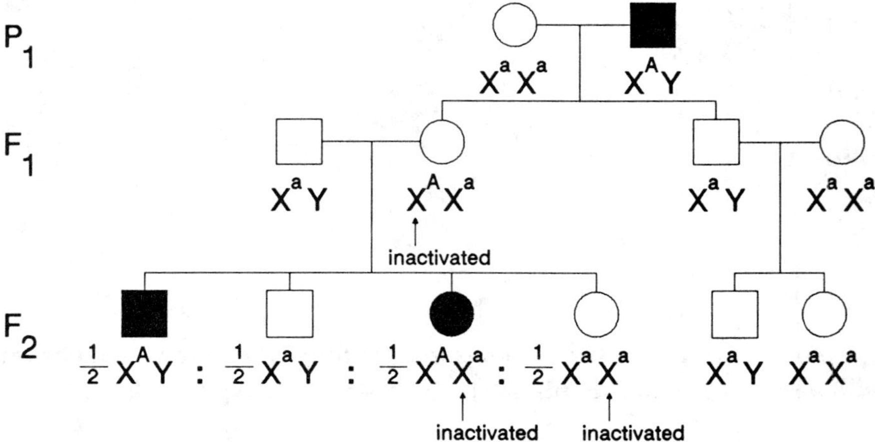

5-3. Answer question 5-2 above for a rare sex-linked <u>recessive</u> trait.

There would be no change in the expression of the trait due to dominance or recessiveness, since a hemizygous condition occurs in both sexes. In this case pseudodominance is the rule (see the figure below)

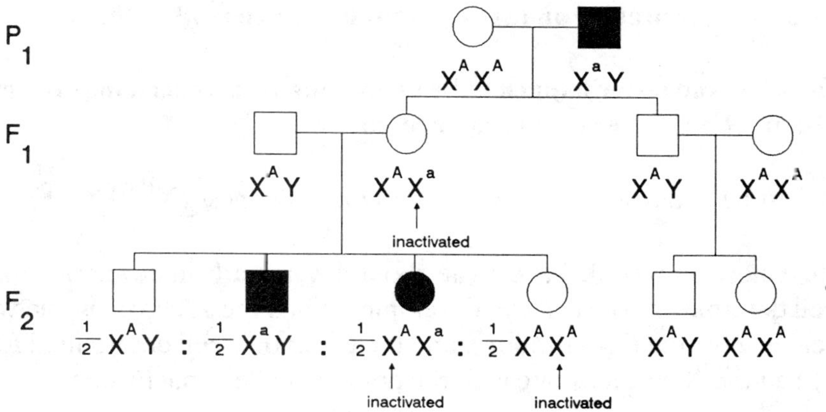

5-4. Is an XYY individual the result of an abnormal complement of chromosomes in the egg, the sperm, or both? At what point in meiosis did this abnormality arise?

An XYY individual is the result of a normal egg with one X chromosome (remember, if your mother had a Y chromosome, she would not be your mother), being fertilized by a YY sperm. The YY gamete is the result of nondisjunction (the failure of chromosomes to separate during cell division) of the Y chromosome during anaphase II of meiosis.

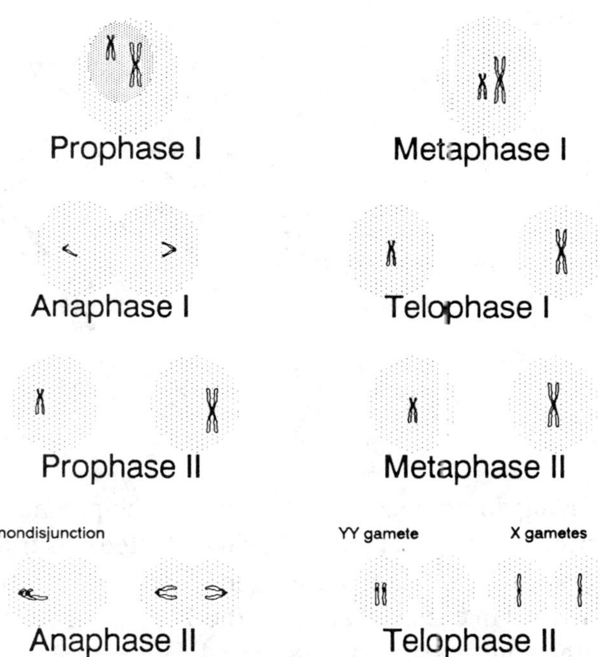

5-5. In humans, red-green color blindness is due to a recessive allele (rsvp) on the X chromosome. A couple, both with normal vision, find out that their child has normal color vision in the right eye, and red-green color blindness in the left eye.

$P \quad X^A X^a \quad X^A Y$

a) What is the sex of the child? Male

b) What would have caused this difference in vision between the right and left eye? Incomplete penetrance

c) What are the genotypes of the parents? $X^A X^a \quad X^A Y$

d) This child has a younger brother and sister. What is the probability that each is color blind?

e) The younger brother and sister happen to be twins. How does this change the probability of one being color blind, if his or her twin is color blind?

a) The child must be a female, heterozygous for color blindness ($X^{Rsvp}X^{rsvp}$).

b) This unusual mosaicism could have come about if, very early in development, the embryonic cell determined for right eye development had the X chromosome with the recessive allele inactivated (Lyonization); and the embryonic cell determined for left eye development had the X chromosome with the dominant allele inactivated.

c) Since both parents are phenotypically normal, the father must have the normal sighted dominant allele on his X chromosome, otherwise he would exhibit the trait. The normal sighted mother must be heterozygous for the gene, passing the recessive allele down to her daughter.

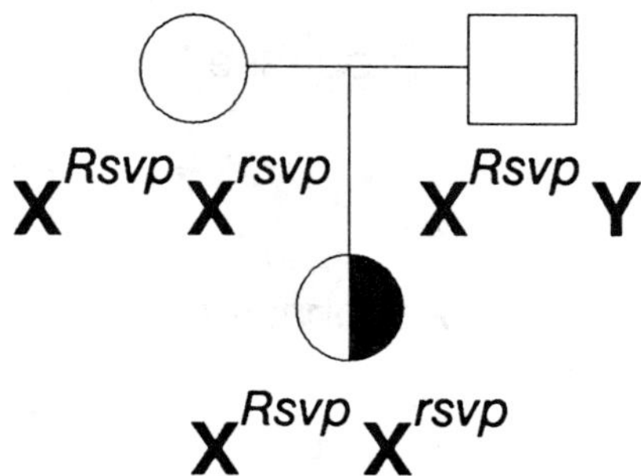

d) Having determined the genotypes of the parents, a Punnett square can be constructed to determine the probability of having affected offspring. The mosaicism exhibited by the first child is extremely rare, because, in humans, one of the X chromosomes is usually randomly inactivated in each cell at about the 12th day of development. There is virtually no chance that the younger sister is color blind. However, there is a 50% chance that the younger brother is color blind.

	X^{Rsvp}	X^{rsvp}	
X^{Rsvp}	$X^{Rsvp}X^{Rsvp}$	$X^{Rsvp}X^{rsvp}$	daughters
Y	$X^{Rsvp}Y$	$X^{rsvp}Y$	sons

e) Since the younger brother and sister are obviously not identical twins (monozygotic twins), the genotype of one has no influence on, and is not indicative of, the genotype of the other.

5-6. Rank the following modes of inheritance based on their frequency of expression of a <u>rare</u> allele in a family pedigree.

Autosomal recessive	_____	**Autosomal dominant**	_____
Sex-linked recessive	_____	**Sex-linked dominant**	_____

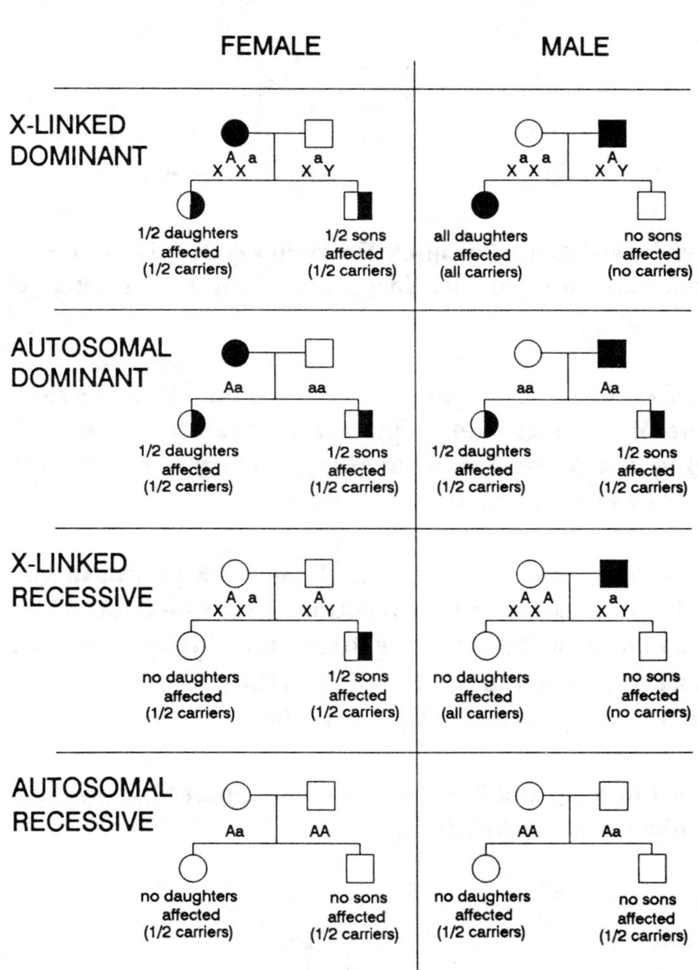

Consider the condition when the mutant allele first appears in an individual male or female. Any trait in a family lineage due to a dominant allele, regardless of sex linkage, would be expressed most frequently because only one allele need be present for the expression of the trait. A sex-linked recessive trait would be expressed next most frequently (given a 1:1 sex ratio of offspring), because 50% of the sons of heterozygous mothers would be affected (due to pseudodominance), and, although none of the children of affected fathers would be affected themselves, all of their daughters would be carriers. Autosomal recessive traits are expressed the least frequent of the modes of inheritance in family pedigrees.

5-7. In chickens, slow feather development (K) is controlled by a dominant, sex-linked allele to rapid feather development (k). Barred plumage (B) is controlled by another sex-linked allele which is dominant to nonbarred plumage (b). Outline both the phenotype and genotype of a cross between a slow, barred hen and a rapid, nonbarred rooster. Continue the lineage by outlining a cross between an F_1 rooster from this mating, and a rapid, nonbarred hen. <u>Hint</u>; remember, chickens are birds.

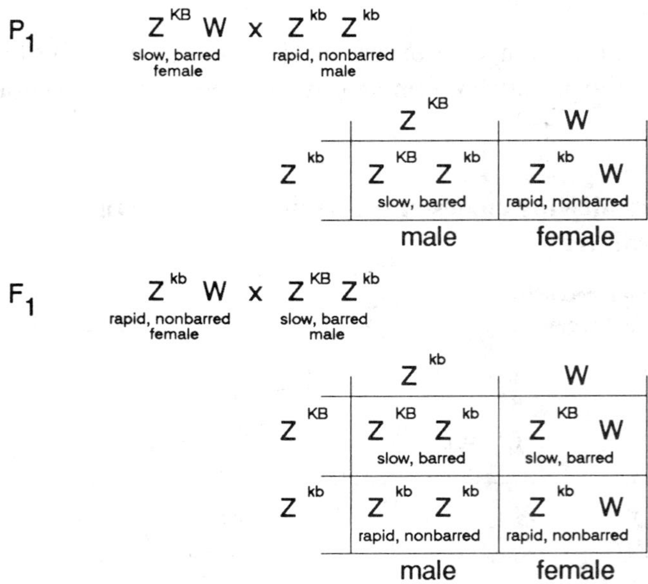

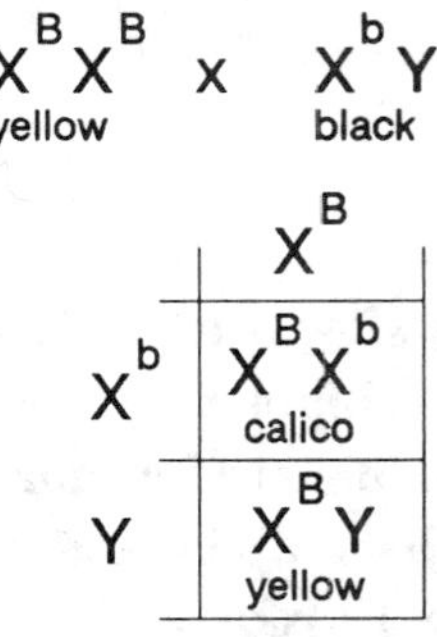

5-8. In cats, coat color is controlled by a sex-linked gene. The allele for yellow coat color (B) is incompletely dominant to the allele for black coat color (b). Heterozygotes exhibit both hair colors, and are referred to as tortoise-shell, or calico cats.

a) Your yellow cat has a litter of four: two calico kittens, and two yellow kittens. You take the kittens to the veterinarian for a check-up. Which sterilization procedure, neutering (the surgical removal of the testes), or spaying (the surgical removal of the ovaries), would the vet recommend for each of the kittens once they were old enough?

b) You decide not to take your vets' advice, and the next year you find one of the calicos has a litter of three: one black, one yellow, and one calico. With the exception of your own cats, there are only two other cats in the area that could be the father of the litter. One of your neighbors has a calico cat, and another has a black cat. Before you start accusing either of your neighbors, how would you narrow down who the potential father could be?

c) You later find out that your neighbors' black cat is a female, which is pregnant from your yellow male. What will be the probable phenotype of the litter.?

a) The calico kittens must be female, since only female cats can be heterozygous for a sex-linked trait. Since the mother is yellow, she must be homozygous dominant (remember heterozygotes are calico), thus contributing the B allele to her calico offspring. The recessive allele (b) of the calicos must have come from the father, a black cat (X^bY). Given this information, the yellow kittens must be males. Therefore the veterinarian would recommend eventual spaying for the calico kittens, and neutering for the yellow kittens.

b) First of all the neighbor with the calico cat is off the hook, since the cat must be a female. To determine if the neighboring black cat is the father, or one of your own yellow males, you would need to determine the sex of the black and yellow kittens (again, the calico kitten is female). If both the black and yellow kittens were males, then your findings would be inconclusive. If, however, one of the black and yellow pair were females (and only one or the other could be!), then the father could be determined.

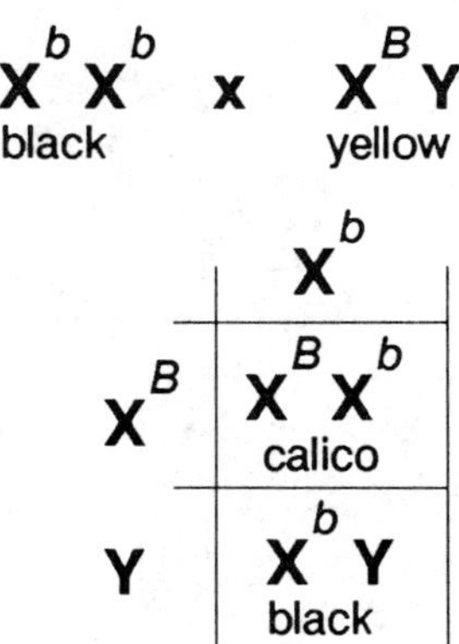

c) All of the females of the litter will be calico. All of the males of the litter will be black.

5-9. Duchenne muscular dystrophy, caused by a recessive X-linked allele (*mdx*), is a disorder involving the gradual deterioration of muscle tissue. Afflicted individuals usually die in their early teens. The disease only develops in boys, never girls. Why?

Afflicted boys receive their recessive allele from their mothers, who are normal, but nonetheless carriers. Afflicted females would have to be homozygous recessive, receiving one of the recessive alleles from a father (who would be hemizygous for the disease, and thus be afflicted). Afflicted males do not survive to reproductive age however, therefore homozygous females never appear in the population.

Identify the possible mode(s) of inheritance for the following pedigrees.

5-10.

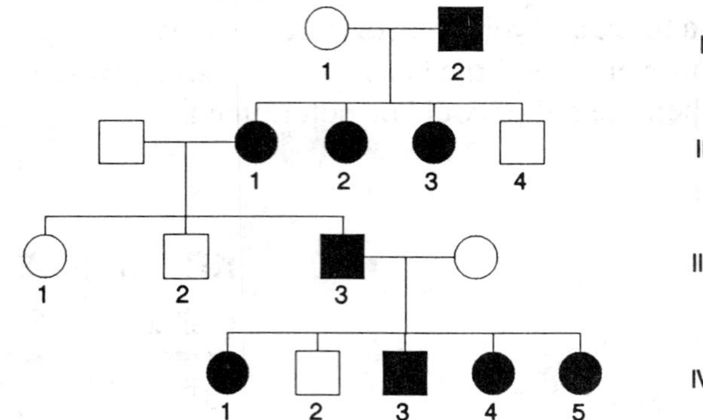

Autosomal dominance - Possible.

Autosomal recessive - Possible, as long as all of the unaffected marriage partners were heterozygotes.

Sex-linked dominance - Not possible, because individual III 3 passed the trait on to one of his sons (IV 3).

Sex-linked recessive - Not possible, because individual III 2 is unaffected.

5-11.

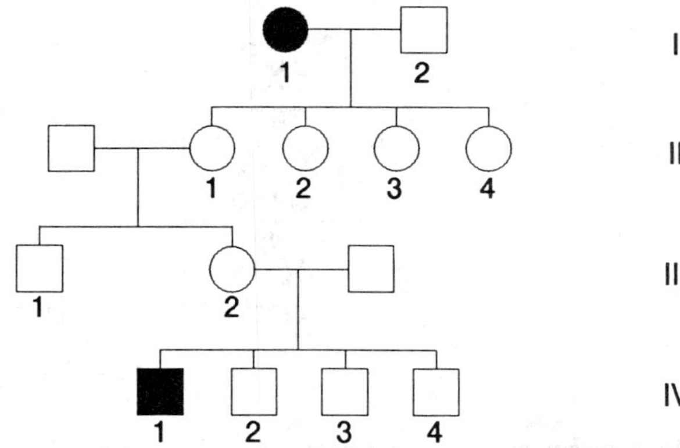

Autosomal dominance - Not possible, because whole generations are not affected.

Autosomal recessive - Possible, if individuals II 1, III 2, and the mate of III 2 were all heterozygous for the trait.

Sex-linked dominance - Not possible, once again because whole generations do not express the trait.

Sex-linked recessive - Possible, again if individuals II 1 and III 2 were carriers (the mate to III 2 need not be a carrier).

5-12.

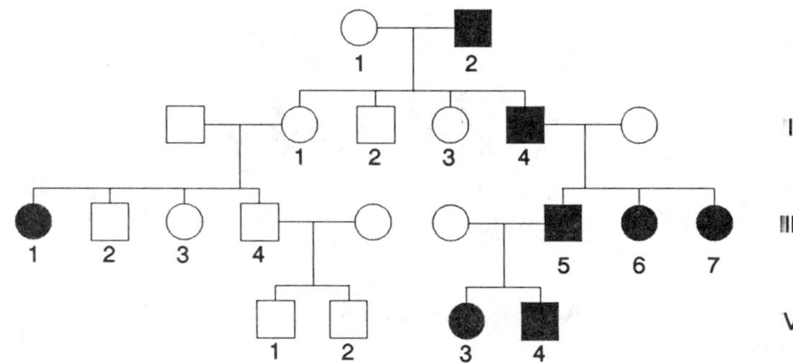

Autosomal dominance - Not possible, because one of the parents of individual III 1 must carry the allele without being affected him/herself.

Autosomal recessive - Possible, individuals I 1, II 1, II 2, III 4, and the mates of individuals II 1, II 4, III 4, and III 5 would have to be heterozygotes.

Sex-linked dominance - Not possible, because the daughters of individual I 2 are unaffected.

Sex-linked recessive - Possible, given that individuals I 1, II 1, and the mates of individuals II 1, II 4, and III 5 are all heterozygotes.

The gene for the enzyme glucose-6-phosphate dehydrogenase (*G6pd*) is located on the X chromosome. There exists two allelic forms of this gene. The *A* form produces an electrophoretically slower polypeptide, and the *B* form produces a more quickly migrating polypeptide. Both of these polypeptides are fully functional protein subunits of the enzyme, which exists as a dimer. Mutant alleles of this gene also exist (*g6pd*) which produce nonfunctional proteins resulting in a glucose-6-phosphate dehydrogenase deficiency.

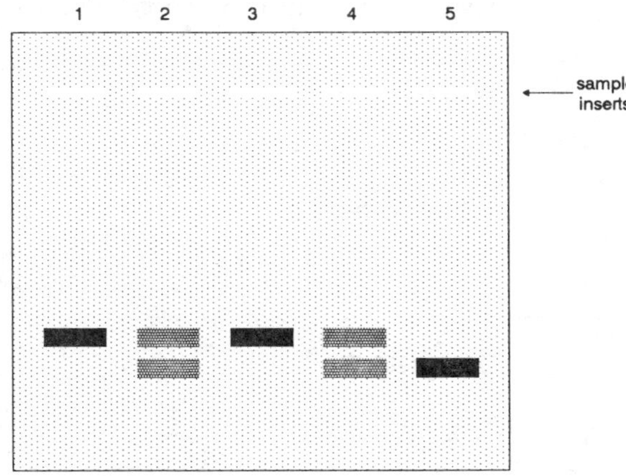

This is a diagram of an electrophoretic gel stained for glucose-6-phosphate dehydrogenase from blood samples of a normal family of five (lanes 1 and 2 are from the parents; lanes 3, 4, 5 are from the children).

5-13. Given the above information, draw a pedigree of this family unit, identifying the sex and genotype of each member.

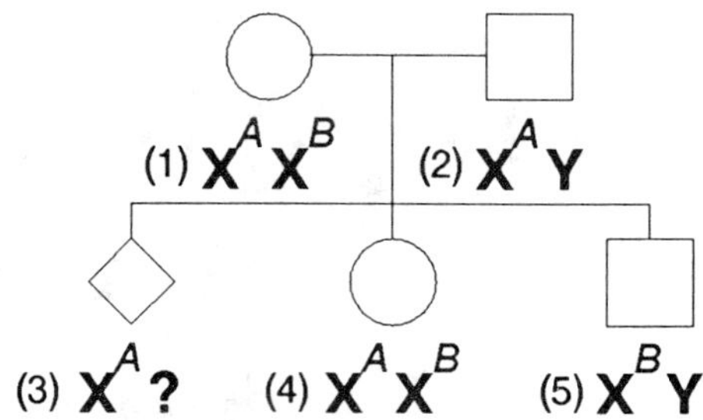

Since the locus is sex-linked, only females can be heterozygous. Therefore, of the parents, lane two must be from the mother. Furthermore, lane 4 must represent a daughter. The father (lane 1) has only the A allele, therefore lane 5 must have come from a son who is hemizygous for the B allele. The sex of the individual represented by lane 3 is uncertain. Both an $X^B X^B$ daughter, or an $X^B Y$ son are possible.

5-14. Are A, B and AB forms of the enzyme glucose-6-phosphate dehydrogenase isozymes, allozymes, or both? Why do you not see an electrophoretic band of the AB form on the gel?

The A, B, and AB forms of glucose-6-phosphate dehydrogenase represent allozymes, since they are different forms of the enzyme controlled by <u>alleles at the same locus</u>. AB forms of the enzyme do not usually occur, because the locus is on the X chromosome. Cells of normal males only possess one X chromosome, and duplicate X chromosomes in female cells (and cells of multiple X males) are normally Lyonized.

5-15. The son of this family marries a woman who has a glucose-6-phosphate dehydrogenase deficiency. What is the probability that their future sons and daughters will also have the deficiency? What are the possible genotypes of their children.

All of their daughters will have the normal B allozyme, and all of their sons will have the *g6pd* deficiency.

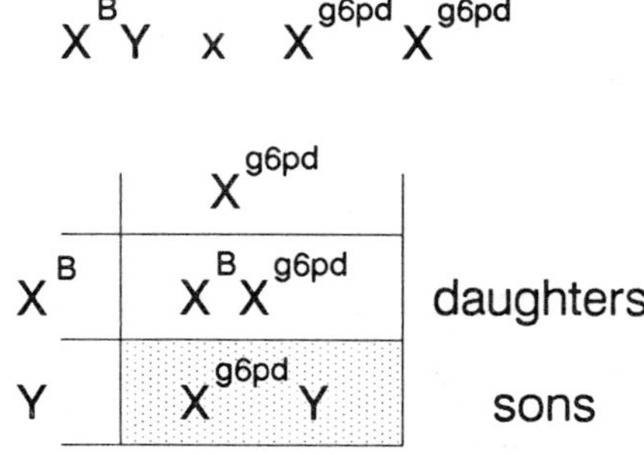

Key Concepts and Terms

Linkage groups	Genome
Nonparental (recombinant)	Parental (nonrecombinant)
Trans configuration (repulsion)	Map unit (centimorgan)
Two-point cross	Cis configuration (coupling)
Reciprocal classes	Three-point cross
Genetic marker	Mapping function
Positive interference	Coefficient of coincidence
Endomitosis	Polytene
Random strand analysis	Chromomeres
Ascospores	Ascus
Ordered spores	Mating types
First division segregation (FDS)	Minimal medium
Unordered spores	Second division segregation (SDS)
Nonparental ditype (NPD)	Parental ditype (PD)
Lod score method (lod, z)	Tetratype (TT)
G-bands	Chromosome banding
Heterokaryon	Somatic-cell hybridization
Synteny test	HAT medium
Clones	Assignment test

Study Questions

6-1. What is the distinction between crossing over and recombination?

Recombination refers to any event that causes variation in the arrangement of allelic forms of genes. Independent assortment during meiosis causes recombination of alleles located on different chromosomes. Crossing over is a specific form of recombination, that of alleles of genes located on the same chromosome.

6-2. Fruit pubescence and fruit length are genetic traits of tomato plants, each controlled by its own gene. The allele for smooth fruit (P) is dominant to the allele for peach appearing fruit (p), and the allele for round fruit (R) is dominant to the allele for elongate fruit (r). In an experiment done in 1928, J. W. MacArthur crossed a true-breeding smooth, long tomato plant with a true-breeding peach round tomato plant. The F_1 was testcrossed, producing an F_2 generation with the following results:

smooth, round	12
smooth, long	123
peach, round	133
peach, long	12

a) Give the genotypes of the P_1 and F_1 crosses.

b) Classify the F$_2$ phenotypes as either parental, nonparental, recombinant, or nonrecombinant products.

c) Determine the map distance between the two loci

```
a)      P1      PPrr   x   ppRR
        F1      PpRr   x   pprr
```

b) The parental, or nonrecombinant, classes can be identified two ways. First, as the name implies, they will have the same phenotypes as the parental generation. Second, they will be the classes that occur with the greatest frequency, since crossing over occurs less frequently than no crossing over. Therefore, the smooth, long and the peach, round classes are the parental (nonrecombinant) classes, and the smooth, round and peach, long classes are the nonparental (recombinant) classes.

F$_2$ phenotype	F$_1$ gamete	numbers	Recombination event p-r
smooth, round	$P\,R$	12	12
smooth, long	$P\,r$	123	-----
peach, round	$p\,R$	133	-----
peach, long	$p\,r$	12	12
total		280	24
percentage			**8.6**

The map distance between the fruit pubescence locus and the fruit length locus is 8.6 map units (or centimorgans)

6-3. A homozygous purple (*pur* - purple eye color), dumpy (*dp* - dumpy wings), black (*bl* - black body) fruit fly is crossed with a pure breeding wild type fruit fly. The F$_1$ are testcrossed, producing the following F$_2$ phenotypic classes:

wild type	756
purple	59
dumpy	414
black	21
purple, dumpy	23
purple, black	417
dumpy, black	60
purple, dumpy, black	750

a) Give the genotypes and phenotypes of the P₁ and F₁ crosses.

b) Determine the gene order of the loci.

c) Rearrange the phenotype classes into reciprocal classes starting with the nonrecombinant classes and ending with the double recombinant classes.

d) Indicate the F₁ gamete contribution to each of the phenotype classes of part c.

e) Calculate the map distances of the loci.

f) Calculate the coefficient of coincidence.

g) Is there positive or negative interference? What is the interference value?

```
a)        P1       purpurdpdpblbl      x      + + + + + +
                 purple, dumpy, black        wild type
          F1       +pur+dp+bl         x      purpurdpdpblbl
                   wild type                 purple, dumpy, black
```

b) The gene order can be revealed by determining which gene arrangement will give the proper double recombinant linkage groups (black and dumpy, purple) given the parental linkage groups (purple, dumpy, black and wild type).

Proposed Gene Order	Double Crossover		genotype	phenotype	Proper Gene Order
pur dp bl / + + +	pur dp bl / + + +		pur + bl / + dp +	purple black dumpy	NO
dp pur bl / + + +	dp put bl / + + +		dp + bl / + pur +	dumpy black purple	NO
dp bl pur / + + +	dp bl pur / + + +		dp + pur / + bl +	dumpy purple black	YES

the *dp-bl* map distance is 35 map units,
the *bl-pur* map distance is 6.5 map units,
and the **best** *dp-pur* map distance is 41.5 map units (35 + 6.5).

F_2 phenotype		F_1 gamete	numbers	Recombination event		
				dp-bl	*bl-pur*	*dp-pur*
wild type		+ + +	756	-----	-----	-----
dumpy, black, purple		*dp bl pur*	750	-----	-----	-----
dumpy		*dp* + +	414	414	-----	414
black, purple		+ *bl pur*	417	417	-----	417
purple		+ + *pur*	59	-----	59	59
dumpy, black		*dp bl* +	60	-----	60	60
black		+ *bl* +	21	21	21	-----
dumpy, purple		*dp* + *pur*	23	23	23	-----
total			2,500	875	163	950
percentage				**35**	**6.5**	**38**

f) Coefficient of coincidence is defined as;

```
(observed double crossovers)/(expected double crossovers)
          (21 +23)/(0.35)(0.065)(2500)
                44/56.9
                 0.77
```

g) Since there are fewer observed double crossover events than expected (coefficient of coincidence 1.0), there is positive interference. The interference value = (1.0 - 0.77) = **0.23**.

6-4. A pure breeding nonbelted, wavy-haired, himalayan mouse is mated with a pure breeding belted, straight-haired, albino mouse to produce an F₁ generation that are all phenotypically nonbelted, wavy-haired, himalayan. The F₁ are testcrossed, producing the following F₂ phenotypes in the corresponding quantities:

nonbelted, wavy, himalayan	33
nonbelted, wavy, albino	30
nonbelted, straight, himalayan	5
nonbelted, straight, albino	6
belted, wavy, himalayan	6
belted, wavy, albino	6
belted, straight, himalayan	32
belted, straight, albino	30

a) Give the genotypes and phenotypes of the P₁ and F₁ crosses.

b) Determine the gene order of the loci.

c) Rearrange the phenotype classes into reciprocal classes starting with the nonrecombinant classes and ending with the double recombinant classes.

d) Indicate the F₁ gamete contribution to each of the phenotype classes of part c.

e) Calculate the map distances of the loci.

f) Calculate the coefficient of coincidence.

g) Is there positive or negative interference? What is the interference value?

a) First, since the F_1 phenotypes are all nonbelted, wavy, himalayan, then these must be controlled by dominant alleles. Let us designate these alleles as $bt+$, Wv, and $a+$ respectively, and the recessive alleles as bt, $Wv+$, and a (if you are unsure why we use these designations reread the nomenclature section of chapter 2 of the Tamarin text). The P_1 and F_1 crosses are therefore;

```
P1          bt+bt+WvWva+a+        x        btbtbWv+Wv+aa
      nonbelted, wavy, himalayan     belted, straight, albino
F1          bt+btWvWv+a+a          x        btbtWv+Wv+aa
      nonbelted, wavy, himalayan     belted, straight, albino
```

b) This is a trick problem because four of the phenotypic classes are of equally high frequency, and the other four are of equally low frequency. The equal frequency of these two groups of four indicates that one of the loci is assorting independently (unlinked). Of the high frequency classes; nonbelted, wavy, himalayan and belted, straight, albino are the parental reciprocal classes. The other two high frequency reciprocal classes (nonbelted, wavy, albino and belted, straight, himalayan) are due to the unlinked locus. Close examination of these two pair of reciprocal classes reveals that they differ by the hair color locus (himalayan vs. albino), indicating that it is the unlinked locus. We are therefore dealing with the following chromosomal arrangement.

```
P1         bt+Wv/bt+Wv a+/a+       x        btWv+/btbWv+ a/a
      nonbelted, wavy, himalayan     belted, straight, albino
F1         bt+Wv/btWv+ a+/a         x        btWv+/btWv+ a/a
      nonbelted, wavy, himalayan     belted, straight, albino
```

c,d,e)

F_2 phenotype	F_1 gamete	numbers	Recombination event bt-Wv	Wv-a	bt-a
nonbelted, wavy, himalayan	+ Wv +	33	-----	-----	-----
belted, straight, albino	bt + a	30	-----	-----	-----
nonbelted, wavy, albino	+ Wv a	30	-----	30	30
belted, straight, himalayan	bt + +	32	-----	32	32
nonbelted, straight, himalayan	+ + +	5	5	UNLINKED 5	
belted, wavy, albino	bt Wv a	6	6	6	-----
nonbelted, straight, albino	+ + a	6	6	-----	6
belted, wavy, himalayan	bt Wv +	6	6	-----	6
total		119	23	73	74
percentage			19.3	61.3	62.2

The *bt-Wv* map distance is 19.3 map units, and both are on different chromosomes from the *a* locus.

f-g) Since there are only two linked loci, there is no way to observe double crossovers. There is no coefficient of coincidence or interference value to calculate.

6-5. In corn; aleurone color (C) is dominant to no aleurone color, or white (c), full endosperm (Sh) is dominant to shrunken endosperm (sh), and starchy endosperm (Wx) is dominant to waxy endosperm (wx). F$_1$ trihybrids were testcrossed and the following phenotypes were recorded in the corresponding quantities:

colored, full, starchy	4
colored, full, waxy	113
colored, shrunken, starchy	2,538
colored, shrunken, waxy	601
white, full, starchy	626
white, full, waxy	2,708
white, shrunken, starchy	116
white, shrunken, waxy	2

a) Give the genotypes and phenotypes of the P$_1$ and F$_1$ crosses.

b) Determine the gene order of the loci.

c) Rearrange the phenotype classes into reciprocal classes starting with the nonrecombinant classes and ending with the double recombinant classes.

d) Indicate the F$_1$ gamete contribution to each of the phenotype classes of part c.

e) Calculate the map distances of the loci.

f) Calculate the coefficient of coincidence.

g) Is there positive or negative interference? What is the interference value?

```
a)      P1          CCshshWxWx          x          ccShShwxwx
          colored, shrunken, starchy       white, full, waxy
        F1          CcShshWxwx          x          ccshshwxwx
          colored, full, starchy       white, shrunken, waxy
```

b) The gene order can be revealed by determining which gene arrangement will give the proper double recombinant linkage groups (colored, full, starchy and white, shrunken, waxy) given the parental linkage groups (colored, shrunken, starchy and white, full, waxy).

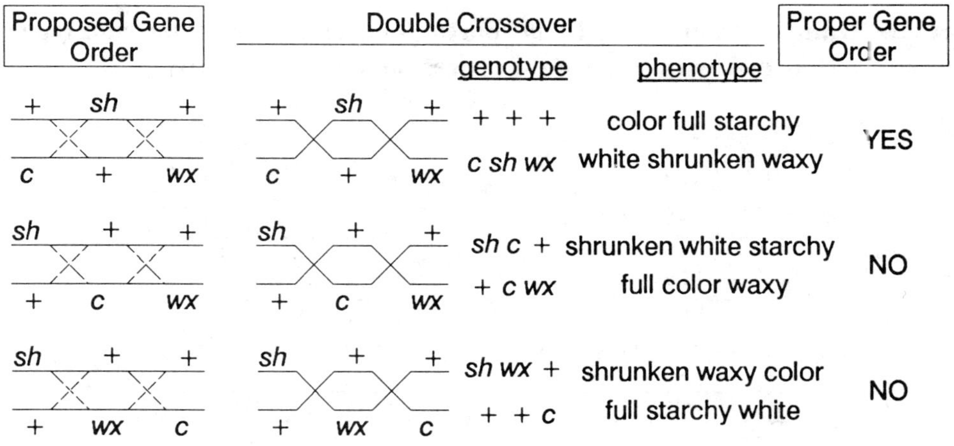

Proposed Gene Order	Double Crossover genotype / phenotype	Proper Gene Order
+ sh + / c + wx	+ sh + / c + wx → + + + color full starchy / c sh wx white shrunken waxy	YES
sh + + / + c wx	sh + + / + c wx → sh c + shrunken white starchy / + c wx full color waxy	NO
sh + + / + wx c	sh + + / + wx c → sh wx + shrunken waxy color / + + c full starchy white	NO

F$_2$ phenotype	F$_1$ gamete	numbers	Recombination event		
			c-sh	sh-wx	c-wx
colored, shrunken, starchy	+ sh +	2,538	-----	-----	-----
white, full, waxy	c + wx	2,708	-----	-----	-----
colored, full, waxy	+ + wx	113	113	-----	113
white, shrunken, starchy	s sh +	116	116	-----	116
colored, shrunken, waxy	+ sh wx	601	-----	601	601
white, full, starchy	c + +	626	-----	626	626
colored, full, starchy	+ + +	4	4	4	-----
white, shrunken, waxy	c sh wx	2	2	2	-----
total		6,708	235	1,233	1,356
percentage			**3.5**	**18.4**	**20.2**

c,d,e)

the *c-sh* map distance is 3.5 map units,
the *sh-wx* map distance is 18.4 map units,
and the **best** *c-wx* map distance is 21.9 map units (3.5 + 18.4).

f) Coefficient of coincidence is defined as;

```
(observed double crossovers)/(expected double crossovers)
     (4 + 2)/(0.035)(0.184)(6708)
          6/43
          0.14
```

g) Since there are fewer observed double crossover events than expected (coefficient of coincidence 1.0), there is positive interference. The interference value = (1.0 - 0.14) = **0.86.**

6-6. In *Neurospora*, a strain which produces fuzzy spores (*fz*) is crossed with the wild type (*fz*$^+$). The following asci were recorded, with some of the spores not developing fully.

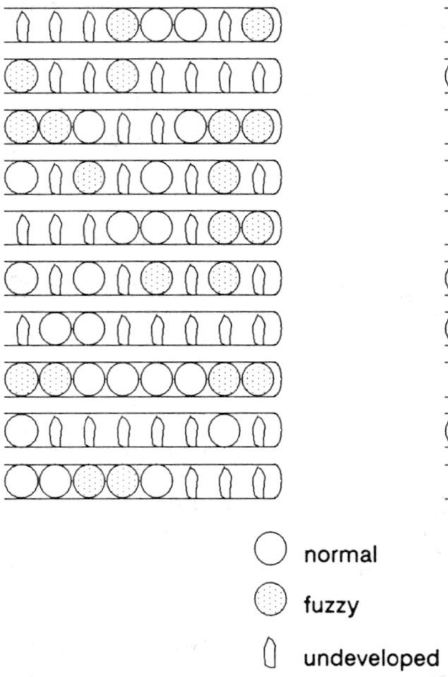

○ normal

◉ fuzzy

◊ undeveloped

a) Determine the phenotype of all of the undeveloped spores, and classify each ascus as to segregational type.

b) Calculate the map distance of the fuzzy locus to the centromere.

a.)

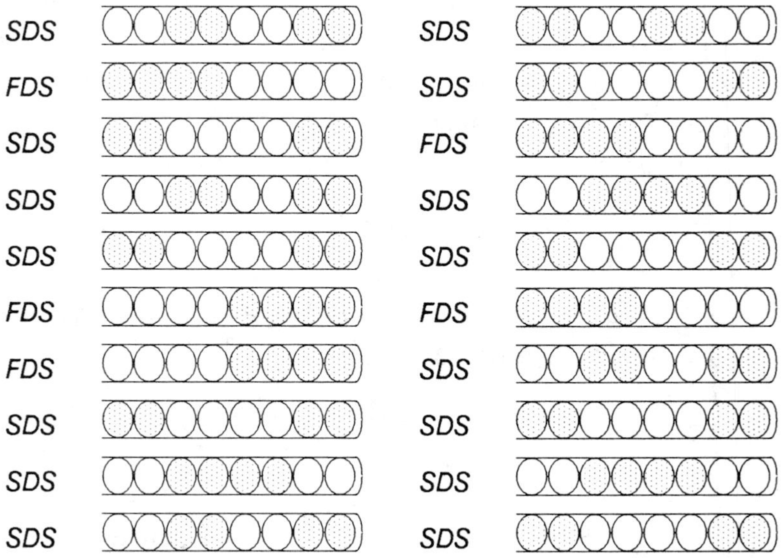

SDS		SDS
FDS		SDS
SDS		FDS
SDS		SDS
SDS		SDS
FDS		FDS
FDS		SDS
SDS		SDS
SDS		SDS
SDS		SDS

b.)

$$\text{Map distance} \; = \; \frac{1/2 \; (SDS)}{\text{total \# of asci}} \; (100)$$

$$= \; \frac{1/2 \; (15)}{20} \; (100)$$

$$= \; 37.5$$

A wild type *Neurospora* strain was crossed with a double mutant strain, *alsm* (albino and small spore size). Ascus development was arrested before the final mitotic division in the resultant colony with the following results:

Ascus type

1	2	3	4	5	6	7
++	+sm	+sm	+sm	++	alsm	+sm
++	+sm	++	alsm	alsm	-+	al+
alsm	al+	alsm	++	++	-sm	+sm
alsm	al+	al+	al+	alsm	al+	al+
392	4	25	50	4	7	5

6-7. Determine the map distance of the *al* and *sm* locus to its centromere.

$$al \;\; \text{map distance} \; = \; \frac{1/2 \; (SDS)}{\text{total \# of asci}} \; (100)$$

$$= \; \frac{1/2 \; (50 + 4 + 7 + 5)}{487} \; (100)$$

$$= \; 6.7$$

$$sm \;\; \text{map distance} \; = \; \frac{1/2 \; (SDS)}{\text{total \# of asci}} \; (100)$$

$$= \; \frac{1/2 \; (25 + 4 + 7 + 5)}{487} \; (100)$$

$$= \; 4.2$$

6-8. Are the two loci linked? Draw the chromosome map of the loci in relation to each other and to their centromere(s).

To determine if the two loci are located on the same chromosome, determine if the loci segregate together more frequently than 50% of the time (the frequency due to independent assortment), or do the allelic rearrangements occur with equal frequency.

The *alsm, alsm, ++, ++* arrangement occurs 392/487 or 80.5% of the time, The *al+, al+, +sm, +sm* arrangement occurs 4/487 or 0.8% of the time. The loci must be linked.

To determine the gene order in relation to the centromere, determine if the *al* locus (the locus farthest from the centromere exhibits crossing over in conjunction with crossing over of the *sm* ("inner") locus. SDS of the *sm* locus are represented by ascus class 3, 5, 6, and 7 for a total of 41 asci. Of those, class 5, 6, and 7 represent SDS of the *al* locus for a total of 16 asci out of 66 *al* SDS's.

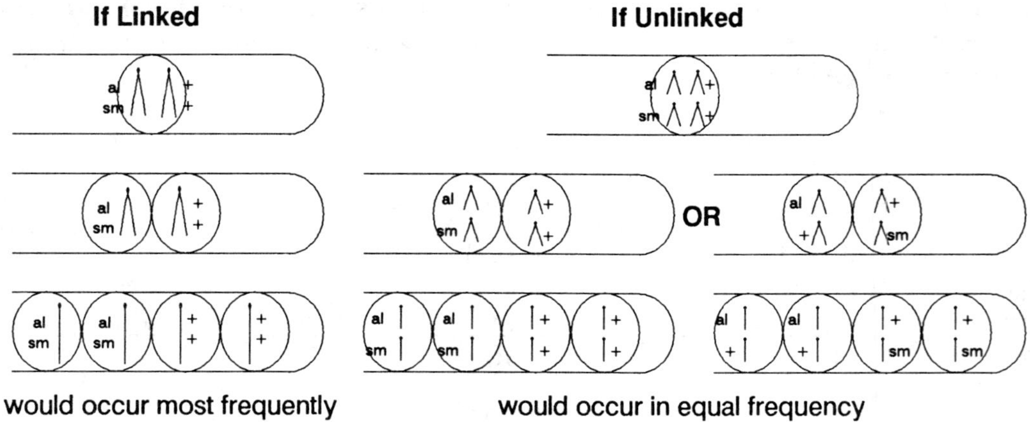

It seems there is no correlation between crossover events between the centromere and the *sm* locus, and crossover event between the centromere and the *al* locus. The chromosomal arrangement must be as follows.

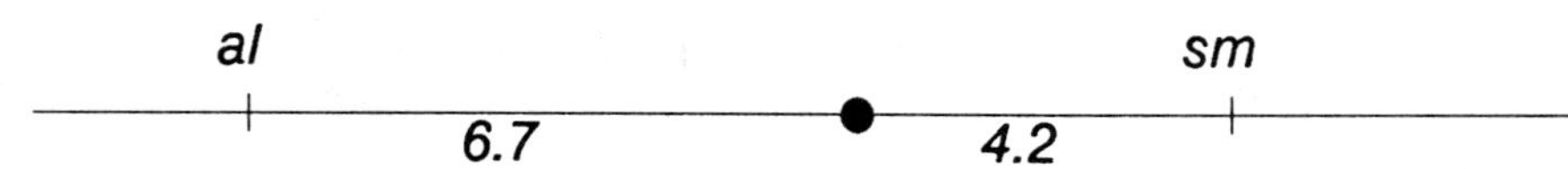

6-9. A strain of yeast that is temperature sensitive (*t*) and requires arginine (*a*) is fused with the wild type. Twenty asci from the resulting diploid colony are dissected and the spores classified as follows (some spores were not viable).

```
1.  a+ at +t            11.      at at ++
2.  a+ at              12. at at ++ ++
3.      at +t ++       13.      ++ ++
4.  at ++ at           14. a+      a+
5.  a+          ++     15.      at at
6.  ++      ++ at      16.      ++ ++ at
7.      +t      +t     17. at at ++ ++
8.  +t ++              18.      ++      ++
9.  ++ ++              19. at at      ++
10. at at ++ ++        20. a+      +t
```

a) Identify the phenotypes of the missing spores, and classify each ascus as to segregational type.

b) Are the two loci linked?

c) If the loci are unlinked, determine the map distance of each to its centromere. If the loci are linked, determine the map distance between them.

```
a) TT    1. a+ at +t ++       PD   11. ++ at at ++
   TT    2. a+ at +t ++       PD   12. at at ++ ++
   TT    3. a+ at +t ++       PD   13. at ++ ++ at
   PD    4. at ++ at ++      NPD   14. a+ +t a+ +t
   TT    5. a+ at +t ++       PD   15. ++ ++ at at
   PD    6. ++ at ++ at       PD   16. at ++ ++ at
  NPD    7. a+ +t a+ +t       PD   17. at at ++ ++
   TT    8. +t ++ at a+       PD   18. at ++ at ++
   PD    9. ++ ++ at at       PD   19. at at ++ ++
   PD   10. at at ++ ++       TT   20. a+ at +t ++
```

b) The loci appear to be linked, because in most of the asci the alleles segregated together (producing parental ditypes).

c) The distance of the loci from the centromere can not be determined because the spores do not remain ordered according the their meiotic division as in *Neurospora*. The map distance between the loci is calculated as follows;

$$\text{Map distance} = \frac{\frac{1}{2}(TT) + (NPD)}{\text{total \# of asci}} (100)$$

$$= \frac{\frac{1}{2}(6) + (2)}{20} (100)$$

$$= 25$$

6-10. Yeast spores from question 6-9 were grown to produce haploid colonies. From these, a colony that is temperature sensitive is mated with a colony that requires arginine. The resulting diploid organism produced 60 asci of the following segregational types;

```
1.    +t a+ ++ at
2.    at ++ ++ at
3.    a+ a+ +t +t
```

a) Identify the segregational types of the asci above.

b) Given that only 5 asci of type 2 were classified, how many asci of type 1 and type 3 would you expect to count?

```
a)    +t a+ ++ at = tetratype
      at ++ ++ at = nonparental ditype
      a+ a+ +t +t = parental ditype
```

b) Having already determined that the map distance between the loci is 25 (from problem 6-9), and knowing that 5 out of 60 asci are NPD, the expected number of PD and TT can be determined.

$$\text{Map distance} = \frac{1/2\ (TT)\ +\ (NPD)}{\text{total \# of asci}}\ (100)$$

$$25 = \frac{1/2\ (TT)\ +\ (5)}{60}\ (100)$$

$$0.25 = \frac{1/2\ (TT)\ +\ (5)}{60}$$

$$15 = 1/2\ (TT)\ +\ (5)$$

$$TT = 20$$

PD = total asci - (NPD + TT)
PD = 60 - 25
PD = 35

6-11. How would strong interference (vs. weak interference) affect a mapping function?

The mapping function is a relationship of measured map distance to actual map distance between two loci. Measured map distance often underestimates actual map distance due to multiple crossovers (even number crossovers) between the loci, and does not exceed 50. Interference is a measure of those factors, whether they are physical or chemical, that hinder multiple crossovers within a region of a chromosome (i.e. between loci). Therefore, under strong interference the measured map distance would more closely reflect the actual map distance.

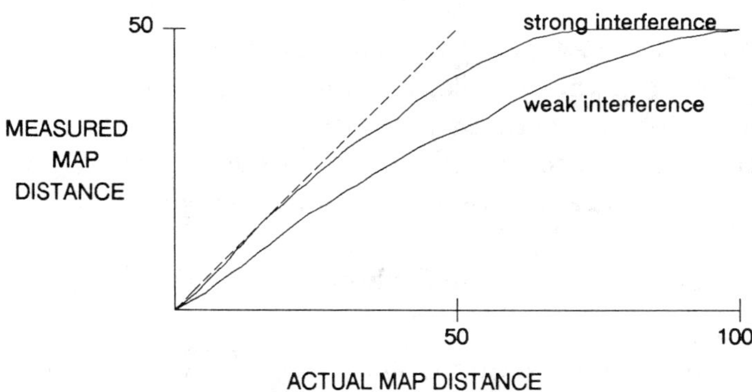

Consider the following scenario. The gene for hypoxanthine phosphoribosyltransferase (*Hprt*) and phosphorylase kinase (*Phk*) are both located on the X chromosome in humans. A young boy is found to have a deficiency for both of these two enzymes (*hprt, phk*).Electrophoretic analysis of his sister shows that, although not affected, she carries both of the mutant alleles. The parents of these siblings, along with a brother, died recently in a car accident, but it is known that none of them exhibited either deficiency. The paternal grandparents and the maternal grandfather are also deceased, but it is known that the maternal grandfather had the phosphorylase kinase deficiency. Upon testing the maternal grandmother, who is still alive and phenotypically normal, it is found that she carries the *hprt* allele.

6-12. Determine whether the mutant alleles are in the cis or trans configuration for the following individuals.

    a) the young boy

    b) his sister

    c) the deceased mother

    d) the maternal grandfather

    e) the maternal grandmother

a) The alleles must be in the cis configuration, since normal males have only one X chromosome.

b) The mutant alleles must be in the cis configuration on the X chromosome contributed by her mother, because it is known that her father exhibited no deficiency from his sole X chromosome.

c) Since the mother was phenotypically normal she would had to have had the dominant alleles (*Hprt__?_Phk__?_*). The *hprt* allele must have come from the grandmother, since the grandfather did not exhibit the deficiency, and the *phk* allele must have come from the affected grandfather, because electrophoretic analysis did not reveal the allele in the grandmother. Therefore, the alleles were in the trans configuration.

d+e) Each of these individuals possessed only one of the mutant alleles.

6-13. Draw the pedigree of the maternal grandparents, parents, and three siblings of this family, including their genotypes and indicating where any recombination event may have occurred.

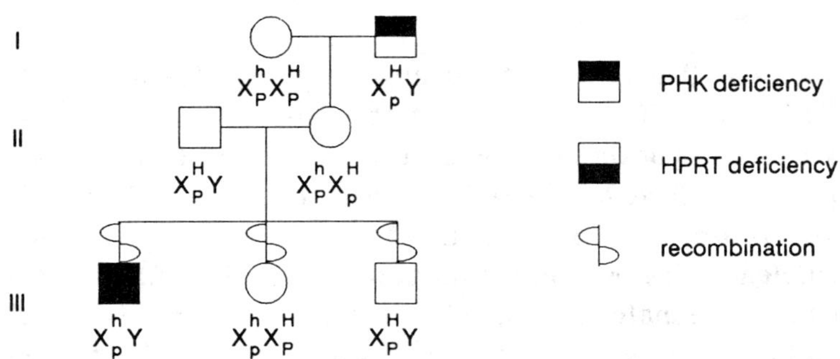

6-14. The following pedigree of rhesus blood type and elliptocytosis (a trait which is controlled by a dominant allele *E*).

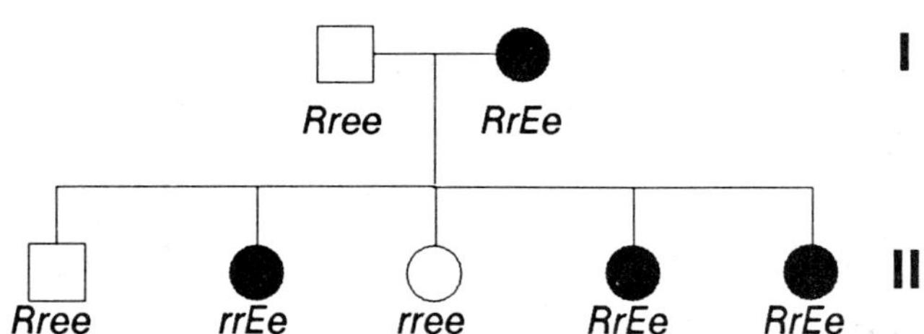

a) Determine a frequency of recombination given that the loci are linked.

b) Test for linkage of the loci for the two traits by calculating a *lod* score

c) Based on the *lod* score do you consider the loci linked?

d) Calculate lod scores for other frequencies of recombination for comparison.

a) Working under the assumption that the loci are linked, there are two possible allelic arrangements; individual I 1 could be *RE/re* (in which case individual II 2 would be a recombinant, *rE/re*), or individual I 1 could be *Re/rE* (in which case individual III 3 would be a recombinant, *re/re*).

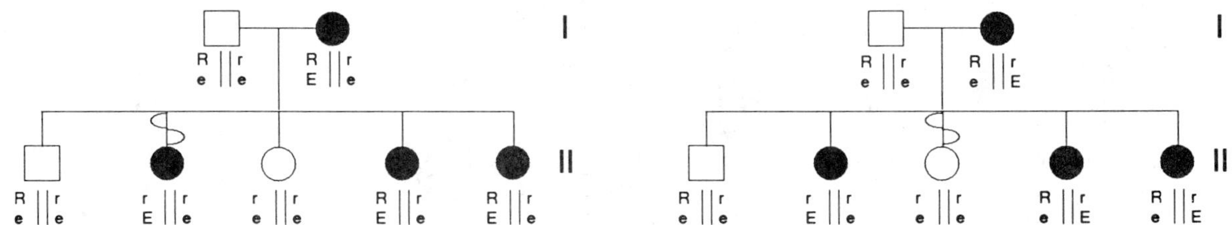

In either of the two linkage arrangements the estimate of map distance would be (1/5)(100) or 20 map units.

b) Assuming the *RE* linkage, the probability of the recombinant child II 2 would be 0.2/2 = 0.1 (divide by 2 because she received one of the two recombination products possible, *rE* and *Re*). The probability of the nonrecombinant children would be (1 - 0.2)/2 = 0.4. If the loci are unlinked, then the genotypic probabilities are based on independent assortment.

RrEe   x Rree

	RE	Re	rE	re
Re	RREe	RRee	RrEe	Rree
re	RrEe	Rree	rrEe	rree

RREe	1/8
RrEe	2/8
RRee	1/8
Rree	2/8
rrEe	1/8
rree	1/8

Our *lod* equation is therefore;

$$z = \text{Log} \frac{(0.1)\,(0.4)^4}{(0.25)(0.125)(0.125)(0.25)\,2}$$

$$= \text{Log} \frac{0.00256}{0.000244}$$

$$= 1.02$$

c) Because the *lod* score is greater than zero, it indicates possible linkage.

d) The following are *lod* scores for other frequencies of recombination. Note that a map distance of 20 produces the best *lod* score.

Map distance	Lod equation	z score
10	$\mathrm{Log}\ \dfrac{(0.05)(0.45)^4}{0.000244}$	0.92
15	$\mathrm{Log}\ \dfrac{(0.075)(0.425)^4}{0.000244}$	1.00
20	$\longrightarrow$	1.02
25	$\mathrm{Log}\ \dfrac{(0.125)(0.375)^4}{0.000244}$	1.01
30	$\mathrm{Log}\ \dfrac{(0.15)(0.35)^4}{0.000244}$	0.96

6-15. Given the table below preform both a synteny and assignment test for each of the human phenotypes under study.

Human-Mouse Hybrid Cell Line	3	5	7	8	10	11	12	13	14	17	18	20	21	22	X	A	B	C	D	E
WIL1				+					+	+			+		+	+	+		+	+
ICL15							+			+		+	+						+	
REX12	+		+			+			+					+	+		+	+		+
XTR2	+	+	+	+		+	+				+	+	+		+	+			+	

A is located on chromosome 8 (those chromosomes common to WIL1 and XTR2 are 8, 21, X, but 21 is excluded because the phenotype did not appear in ICL15, and X is excluded because the phenotype did not appear in REX12).

B and *E* are syntenic and located on chromosome 14.(*B* and *E* show the same pattern of appearance in the hybrid cell lines, WIL1 and REX12. WIL1 and REX12 have only chromosomes 14 and X in common and X can be excluded because there is no phenotypic appearance in cell line XTR2).

C is located on chromosome 7, 11, or 22 (*C* is found in REX12 only. REX12 contains chromosomes 3, 7, 11, 14, 22, and X. Of these chromosomes 3, 14, and X can be eliminated because other cell lines contain these chromosomes but do not show the phenotype.

D is located on chromosome 21 (*D* appears in cell lines WIL1, ICL15, and XTR2. The only chromosome these three cell lines have in common is 21.

Key Concepts and Terms

Bacillus	Coccus
Bacteriophage (phage)	Spirillum
Capsomeres	Capsid
Autotroph	Heterotroph
Auxotroph	Synthetic medium
Conditional-lethal mutant	Prototroph
Complete medium	Enriched medium
Selective medium	Minimal medium
Lysis	Bacterial lawn
Replica plating	Plaques
Virion	Screening technique
Genome	Genophore
Competence factor	Transformation (transformation mapping)
F factor (fertility factor, F^+, F^-)	Conjugation
Hfr (high frequency recombination)	Plasmid
F-pili (sex pili)	Pili (fimbriae)
Exogenote	Merozygote
Interrupted mating	Endogenote
Lysate	Sexduction (F-duction, F')
Lysogenic	Lysogeny
Prophage	Temperate
Transduction	Induction (zygotic induction)
Generalized transduction	Specialized transduction (restricted)
Cotransduced	Transducing particle

Study Questions

7-1. Distinguish between heterotrophic, autotrophic, prototrophic, and auxotrophic organisms.

Heterotrophic and autotrophic identify an organism by the form of carbon (organic, or CO_2 respectively) it utilizes. Prototrophic and auxotrophic identify organisms as being wild type, or genetically deficient in some biochemical pathway. The two classes of terms are not mutually exclusive of each other (i.e. a bacterial strain can be an auxotrophic autotroph).

7-2. How many genetic markers are needed to detect bacterial recombinants in conjugation experiments? Describe the function of the marker(s) and provide an example.

A minimum of two genetic markers are necessary in bacterial recombination studies. One locus must act as a selective marker, to distinguish the F^+, or donor strain, from the recipient. A selective condition is set up based on the marker phenotype to prevent the donor strain from growing on the particular selective media, while allowing the growth of the recipient strain. A second genetic marker is used to identify that recombination has occurred, that is, that the marker phenotype of the donor strain is observable in the recipient strain growing on the selective media.

Common markers include genetic mutations which involve: the inability to utilize particular carbon sources (i.e. sugars); biosynthetic pathways, making them nonfunctional and causing the strain to require the presence of a particular nutrient (i.e. amino acids); or sensitivity (or resistance) to environmental factors (i.e. antibiotics, poisons, temperature). For example, in a cross between Hfr(met^+str^s) and F⁻(met^-str^r), conjugants could be plated on a medium (enriched with methionine) containing streptomycin. This would eliminate all of the donor cells. The resulting colonies could then be plated onto medium without methionine. Subsequent colony growth would indicate that the recipient strain had acquired the ability to synthesize methionine through recombination with the donor strain.

7-3. Briefly outline the steps of bacterial transformation.

1. The binding of exogenous DNA to the cell membrane.
2. The penetration of the DNA into the cell.
3. The synapsis of the foreign DNA to the host chromosome.
4. The integration of the foreign DNA into the host chromosome.

7-4. What specific attributes and events must be present for each of the steps of transformation outlined in 7-3 to occur?

1. The cells must have the competence factor necessary to bind the exogenous DNA to the cell membrane.
2. The exogenous DNA must be double stranded. The energy released from the degradation of one of the strands is used to provide for the penetration of the DNA through the cell membrane.
3. Chromosomal homology facilitates synapsis, although it is not required for integration.
4. Two crossover events must occur to incorporate the foreign DNA into the host chromosome.

7-5. For each of the following terms; identify it as either a phenotypic or genotypic designation, describe the biochemical condition represented, and state the conditions which would favor growth (the first is done as an example).

met-;   genotype; unable to synthesize methionine; requires methionine to grow.
thr-
ala+
Pro+
xyl-
Pens
Ile-
his-lys-
Man-Trp-
tonAr
azirstrsgal-thy-malK+

met-;   genotype; unable to synthesize methionine; requires methionine to grow.

thr-;   genotype; unable to synthesize threonine; requires threonine to grow.

ala+;   genotype; able to synthesize alanine; does not require alanine to grow.

Pro+;   phenotype; able to synthesize proline; does not require proline to grow.

xyl-;   genotype; unable to utilize xylose as a carbon source; requires a carbon source other than xylose to grow.

Pens;   phenotype; sensitive to penicillin; will grow in the absence of penicillin.

Ile-;   phenotype; unable to synthesize isoleucine; requires isoleucine to grow.

his-lys-;   genotype; unable to synthesize histidine or lysine; requires both histidine and lysine to grow.

Man-Trp-;   phenotype; unable to utilize manose as a carbon source and unable to synthesize tryptophan; requires a carbon source other than manose, and requires the presence of tryptophan to grow.

tonAr;   genotype; resistant to T1 and T5 phage infections; able to grow in the presence of T1 and T5 phage.

azirstrsgal-thy-malK+; genotype; resistant to sodium azide, sensitive to streptomycin, cannot utilize galactose as a carbon source but can utilize maltose, cannot synthesize thymine; able to grow on a medium containing sodium azide and utilize maltose as a carbon source as long as thymine is also present (cannot grow on galactose, or in the presence of streptomycin).

7-6. Two auxotrophic strains of bacteria are allowed to conjugate in complete liquid media for a specified period of time. One strain is a Hfr ($azi^s ala^- gly^- val^-$), the other is F⁻ ($azi^r arg^- leu^- lys^-$). A sample of the culture is appropriately diluted and plated on complete medium containing sodium azide. The subsequent colonies are replica plated onto the following selective media. Label the colonies 1-10, identify the phenotype of each, and indicate the genetic contribution of the donor strain which was successfully integrated into the recipient strain.

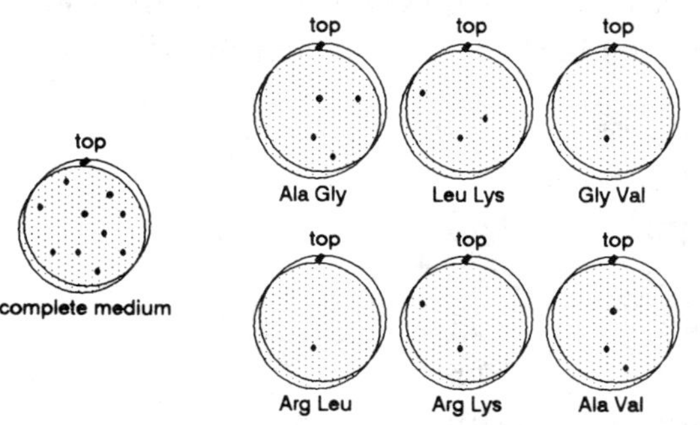

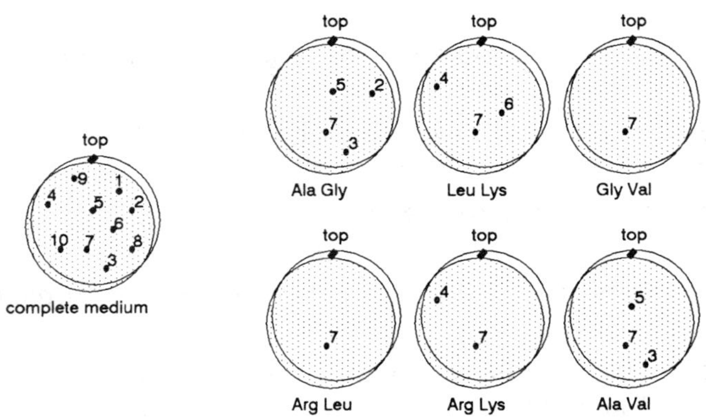

Colony

1,8,9,10	Azi^{r}Arg$^-$Leu$^-$Lys$^-$, are most probably nonrecombinant colonies identical to the recipient strain receiving no genetic material from the Hfr strain.
2	Ala$^-$Gly$^-$, integrating $arg^+leu^+lys^+ala^-gly^-$ from the Hfr.
3	Ala$^-$, integrating $arg^+leu^+lys^+ala^-$ from the Hfr.
4	Lys$^-$, integrating arg^+leu^+ from the Hfr.
5	Ala$^-$, integrating $arg^+leu^+lys^+ala^-$ from the Hfr.
6	Leu$^-$Lys$^-$, integrating arg^+ from the Hfr.
7	wild type, integrating $arg^+leu^+lys^+$ from the Hfr.

7-7. DNA from a prototrophic strain of _Bacillus subtillus_ was used in an experiment to transform a competent auxotrophic strain that was unable to synthesize serine (_ser$^-$_) and threonine (_thr$^-$_). Colonies of the following transformant classes were produced in the quantities specified.

```
ser+ thr+      850
ser+ thr-      217
ser- thr+      205
```

a) What is the frequency of crossing over between the _ser_ and _thr_ loci?
b) What is the map distance between the two loci?

a) The frequency of crossovers between the loci is (217+205)/1272 = 0.332.
b) The percentage of crossover between loci is used as the relative map distance. Therefore the map distance between _ser_ and _thr_ is 33.2.

7-8. DNA from a prototrophic strain of _Bacillus subtillus_ was used to transform a competent auxotrophic strain that was unable to synthesize alanine (_ala$^-$_), arginine (_arg$^-$_), and proline (_pro$^-$_). Transformant colonies were produced in the following quantities.

$ala^+arg^+pro^+$	2100
$ala^+arg^+pro^-$	530
$ala^+arg^-pro^+$	350
$ala^-arg^+pro^+$	110
$ala^+arg^-pro^-$	205
$ala^-arg^+pro^-$	215
$ala^-arg^-pro^+$	208

a) What is the calculated map distance between the three loci?
b) What is the linkage arrangement of the loci?
c) Does the linkage arrangement affect the calculated map distances of part a (if so how and why)?

To determine the map distance of the loci to each other we must first classify the phenotypes as to being single transformants or double transformants of the two particular loci of interest. The ratio of single transformants to double transformants indicates the map distance.

		ala-arg	*arg-pro*	*ala-pro*
$ala^+arg^+pro^+$	2100	double	double	double
$ala^+arg^+pro^-$	530	double	single	single
$ala^+arg^-pro^+$	350	single	single	double
$ala^-arg^+pro^+$	110	single	double	single
$ala^+arg^-pro^-$	205	single	none	single
$ala^-arg^+pro^-$	215	single	single	none
$ala^-arg^-pro^+$	208	none	single	single
single transformants		880	1300	1050
double transformants		3510	3510	3510
		25%	37%	30%

b) Based on the relative map distance, the linkage arrangement must be *pro-ala-arg* (or *arg-ala-pro*).

c) A better estimate of the *pro-arg* map distance is 55 map units (the sum of the *pro-ala* and *ala-arg* distances). The calculated *pro-arg* distance is an underestimate due to multiple crossovers occurring between the loci, which is phenotypically hidden.

Two Hfr strains of *E. coli* that are glu^+ and xyl^- are mated with an F⁻ auxotrophic strain using interrupted mating techniques. After each specified mating interruption, a sample of the culture is plated out on various selective media to determine the phenotypes of various colonies.

The table summarizes the results.

	Time of appearance	
Donor loci	Hfr P801	Hfr G11
argG	64*	3
chlE	14	53
lac	4	63
pabB	36	31
purB	21	46
purC	49	18
purD	--	82
serA	58	9

* note: no colonies appeared after ~75 minutes.

7-9. What carbon source would you have provided on the selective media plates for the conjugants to grow, and why?

The object of all of the selective media plates is to prevent the Hfr donor strain from growing while allowing the resulting conjugant cells to grow. Providing xylose as the sole carbon source would accomplish this objective.

7-10. Based on the table above, construct your own *E. coli* chromosome indicating the following;

the order of the loci,

the distance between the loci in minutes,

the points at which the F factor is located for each

of the strains.

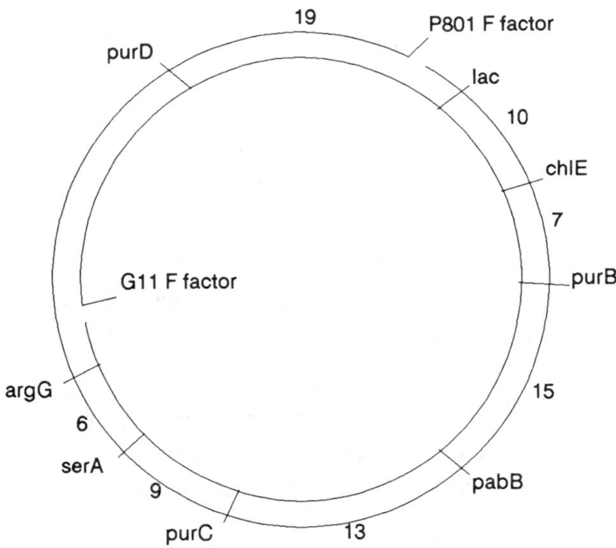

7-11. The transfer of the *purD* locus was routinely absent in the conjugation experiments involving the Hfr P801 strain and as noted no colonies appeared after approximately 75 minutes of conjugation. Provide a possible explanation for these observations. Hint: Although not necessary to answer this question, the chromosomal map of *E. coli* (figure 7.31 of the Tamarin text) provides the answer.

The first explanation one may come up with is that for one reason or another conjugation is interrupted before the transfer of the *purD* locus from the Hfr P801 strain. This would be feasible if the locus were close to the F factor of the chromosome. However another explanation would be that the *xyl* locus is located between the starting point of genetic transfer of the Hfr P801 strain and the *purD* locus.

Under these conditions in order for the *purD* locus to be transferred to the recipient cell the *xyl* locus of the Hfr strain would also have to be transferred and incorporated into the recipient chromosome as well. These conjugants with the Xyl⁻ phenotype would not have grown on the selective media. The notation that no colonies grew after 75 minutes suggests that the *xyl* gene may be located there.

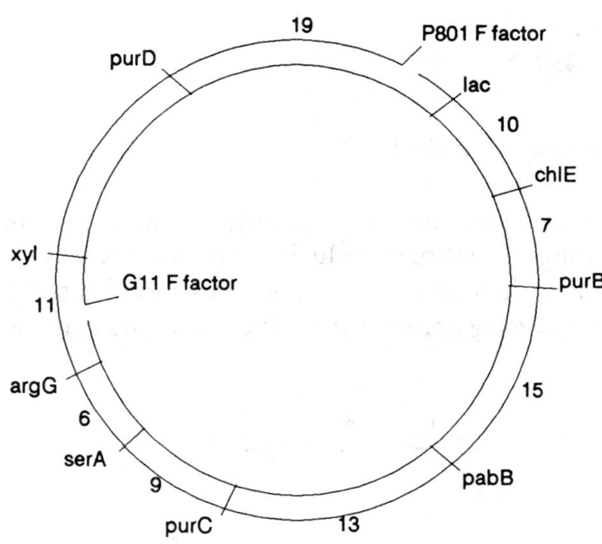

7-12. Rank the following in order of frequency of gene transfer and chromosome integration from one cell to another.

Hfr strain

F⁻ strain

F⁺ plasmid

F' plasmid

1. Hfr strain -   though the F factor is rarely transferred.

2. F' plasmid -   high rate of integration due to a region of chromosome homology.

3. F⁺ plasmid -   1/5 conversions of F⁻ to F⁺.

4. F⁻ strain -   no transfer due to the lack of the fertility factor.

A generalized transducing phage containing 2 auxotrophic mutations (a^- and b^-) and a wild type (c) locus from a bacterial strain is used to infect a second bacterial strain which is wild type for the a and b loci but contains a mutant c locus. Transductant colonies were selected for the c locus with the following results.

1	$a^-b^-c^+$	928
2	$a^+b^-c^+$	2
3	$a^+b^+c^+$	33
4	$a^-b^+c^+$	450

7-13. What is the linkage arrangement of the loci?

The transducing phage contained the $a^-b^-c^+$ genotype which is also the class that occurred most frequently, indicating the entire transducing chromosome was incorporated. The least frequent class ($a^+b^-c^+$) would represent a double crossover occurring within the incorporation of the transducing chromosome. Therefore, the gene order must have the a gene as the middle locus.

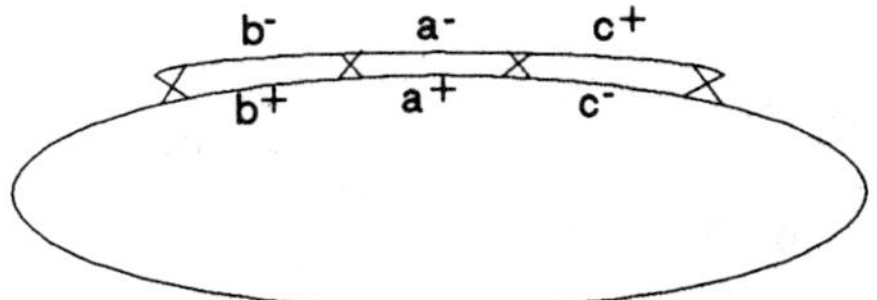

7-14. Calculate cotransductance values to determine the relative distance between the loci. Which two loci are closest?

Cotransductance of a-c = $(928+450)/(928+2+33+450)$ = 0.975
Cotransductance of b-c = $(928+2)/(928+2+33+450)$ = 0.658

The contransductance values indicate that the c locus is much closer to the a locus than it is to the b locus. No reliable information is available concerning the a-b distance because the c locus was the selective marker in determining transductants.

7-15. For each of the classes of transductants draw the crossovers that must have occurred.

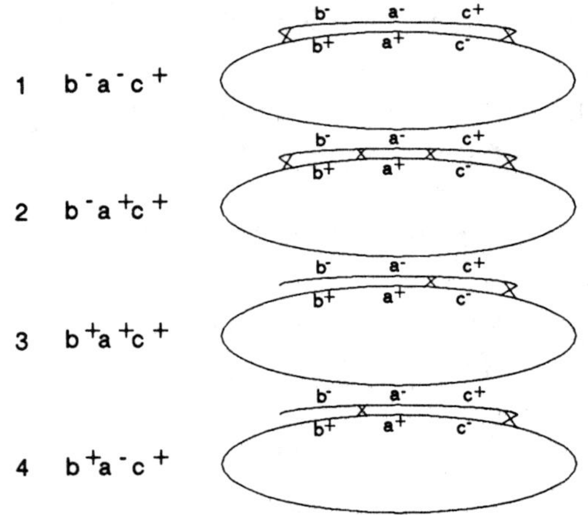

1 $b^-a^-c^+$

2 $b^-a^+c^+$

3 $b^+a^+c^+$

4 $b^+a^-c^+$

Key Concepts and Terms

Centric fragment	Acentric fragment
Dicentric chromosome	Deletion chromosome
Inversion	Breakage-fusion-bridge cycle
Variegation	Position effect
Crossover suppression	Inversion heterozygote
Pericentric inversion	Paracentric inversion
Reciprocal translocation	Supergene
Alternate segregation	Reciprocal translocation heterozygote
Adjacent-2 type segregation	Adjacent-1 type segregation
Fundamental number (NF)	Robertsonian fusion
Unequal crossing over	Centromeric fission
Aneuploidy	Euploidy
Nullisomic	Monosomic
Mosaics (chimeras)	Trisomic
Triploid	Gynandromorph
Polyploid	Tetraploid
Allopolyploidy	Autopolyploidy
Amphidiploid	Somatic doubling

Study Questions

8-1. What typically happens to the genetic information on the acentric fragment of a single break chromosome? In terms of genetic content, how is it different (if at all) from a single break of a chromatid?

An acentric chromosomal fragment does not undergo normal cell division processes and is eventually degraded. The loci of the acentric fragment are lost from the genome of the cell. Single breaks of chromatids do not affect the entire chromosome. Although the acentric fragment is lost, the cell still has the genetic information on the sister chromatid. Only after cell division, and the separation of the sister chromatids, will one of the daughter cells exhibit the deletion effects due to the loss of the acentric fragment.

8-2. List the possible consequences of a chromosomal inversion.

Position effects (variegation)

New linkage groups

Semisterility

Crossover suppression

8-3. By what mechanism does an inversion cause crossover suppression?

Crossover suppression is, in actuality, a misnomer. Inversions do not so much suppress the chromosomal phenomenon of crossing over, but the products of crossover events involving inversions include duplications and deletions. This results in semisterility or loss of viability of the daughter cells. Because of this, the observable effect to the investigator is that crossing over is suppressed.

8-4. What is an inversion heterozygote? Diagram the synapsis of an inversion heterozygote of a tetrad in which the inversion involves 4 of 7 loci and the centromere. Is this a paracentric or pericentric inversion?

Inversion heterozygote refers to the situation in which one chromosome of a homologous pair has a chromosomal inversion and the other does not.

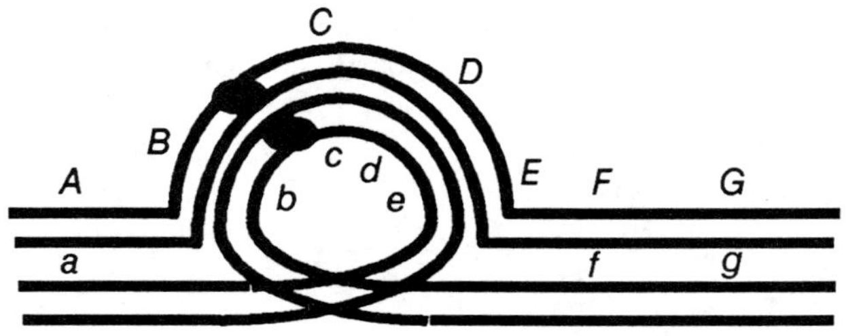

Pericentric inversion

8-5. Show the resulting chromosomal structures if a crossover event were to occur within the inversion loop versus outside the inversion loop.

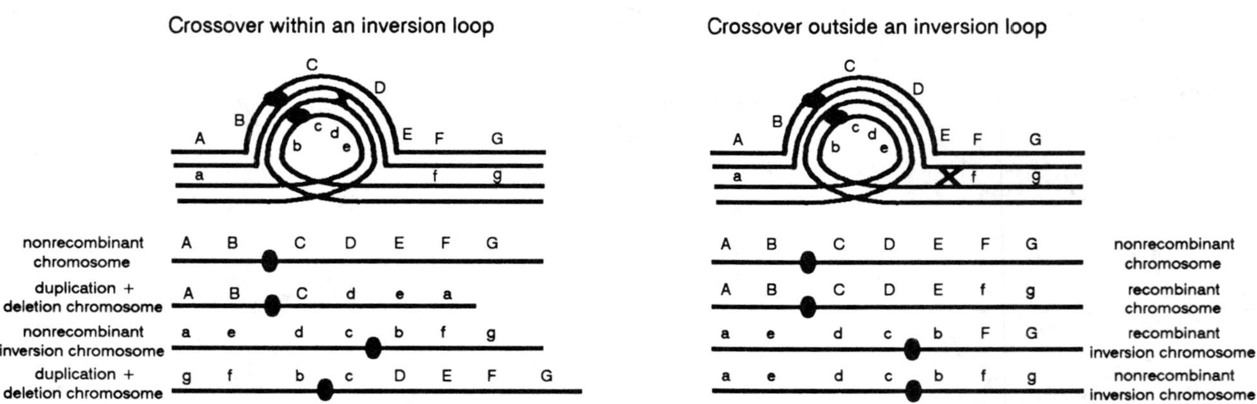

8-6. How do the chromosomal products of a crossover event within a paracentric inversion loop differ from the chromosomal products of a crossover event within a pericentric loop?

Because the centromeres are located outside of the inversion loop in the paracentric configuration, one of the resulting chromosomal structures of the crossover will be a dicentric chromosome and its complementary chromosome product will be an acentric fragment.

8-7. What chromosomal rearrangements will result in new linkage groups and possible position effects?

Deletions
Duplications
Inversions
Translocations
Robertsonian fusions
Centromeric fissions

8-8. List three chromosomal assortment possibilities during meiosis of a reciprocal translocation heterozygote. Which would be observed most frequently, which would you expect to occur least frequently?

1. Alternate segregation would be observed most frequently because the resulting chromosomes would be complete and balanced.

2. Adjacent-1 type segregation is equally likely to occur as alternate segregation, however it results in chromosomes with duplications and deletions that are usually lethal.

3. Adjacent-2 type segregation is least frequent, requiring homologous centromeres to migrate to the same pole during meiosis.

8-9. By what processes can gene duplication arise in gamete formation?

Breakage-fusion-bridge cycle

Inversion loop crossovers

Adjacent-1 and adjacent-2 segregation of a reciprocal translocation heterozygote.

Unequal crossing over

8-10. Provide the clinical designation for the following real and imaginary chromosomal anomalies in humans ($2n = 46$).

Condition	Genetic Designation
Female, trisomic 21	______________
Male, nullisomic 3	______________
Female, monosomic 14	______________
Male, double trisomic 13, 18	______________
Male, translocation of part of the long arm of 2 to the long arm of 10	______________
Female, translocation of the short arm of 8 to the long arm of 21	______________
Klinefelter male, translocation of the short arm of 7 to the short arm of 14	______________
Triploid female	______________
Male, deletion on the short arm of 4	______________

Condition	Genetic Designation
Female, trisomic 21	47,XX,+21
Male, nullisomic 3	44,XY,-3
Female, monosomic 14	45,XX,-14
Male, double trisomic 13, 18	48,XY,+13,+18
Male, translocation of part of the long arm of 2 to the long arm of 10	46,XY,t(2q-;10q+)
Female, translocation of the short arm of 8 to the long arm of 21	46,XX,t(2p-;21q+)
Klinefelter male, translocation of the short arm of 7 to the short arm of 14	47,XXY,t(7p-;14p+)
Triploid female	69,XXX
Male, deletion on the short arm of 4	46,XY,4p-

Key Concepts and Terms

Polygenic (quantitative) inheritance pattern
Mean
Variance (V or s^2)
Parameter
Standard error of the mean (SE)
Covariance
Heritability (H)
True heritability
Heritability in the broad sense (H_B)
Concordance

Continuous (quantitative, metrical) variation
Additive model
Normal distribution
Statistics
Standard deviation (s)
Correlation coefficient (r)
Inbreeding depression
Realized heritability
Heritability in the narrow sense (H_N)

Study Questions

9-1. How do environmental factors affect the phenotypic distribution of a trait that is for the most part under genetic control? Graphically show the difference between a trait that experiences little environmental effects and one that has much environmental control.

Environmental effects on a particular phenotype increase the variation of that trait around an expected mean.

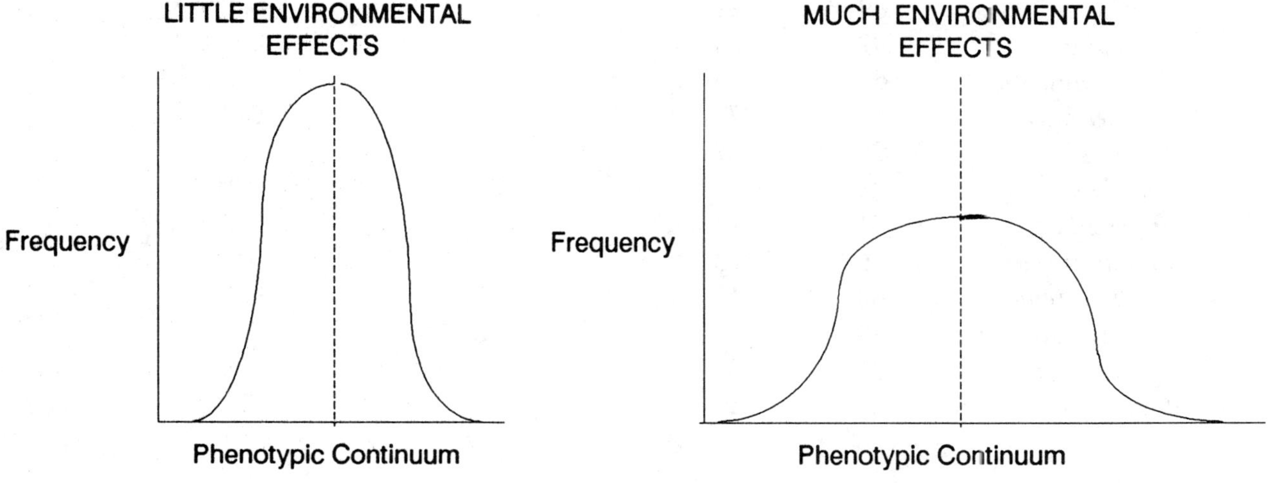

Concordance frequencies were determined in monozygotic twins (MZ) and dizygotic twins (DZ) for the following traits.

	MZ	DZ
Hair color	89	22
Eye color	99.6	28
Blood pressure	63	36
Handedness	79	77
Measles	95	87
Clubfoot	23	2
Tuberculosis	53	22
Mammary cancer	6	3
Schizophrenia	80	13
Down syndrome	89	7
Spina bifida	72	33
Manic-depression	80	20

Which phenotypes would you predict to have high heritability? Which phenotype would you predict to have low heritability?

High concordance of monozygotic twins as compared with dizygotic twins is an indicator of the heritability of a trait. The straight difference between these two values can be used ($C_{MZ} - C_{DZ}$) though some prefer to standardize the difference, assuming a best case rule of 100% concordance for monozygotic twins [$(C_{MZ} - C_{DZ})/(100 - C_{DZ})$]. Depending on which comparative measure you choose to use Down syndrome and eye color appear to be highly heritable, while the heritability of handedness and mammary cancer appear low.

	MZ	DZ	(MZ - DZ)	(MZ - DZ)/(100-DZ)
Hair color	89	22	67 *	0.859 *
Eye color	99.6	28	71.6 *	0.994 **
Blood pressure	63	36	27	0.422
Handedness	79	77	2 **	0.087 *
Measles	95	87	8	0.615
Clubfoot	23	2	21	0.214
Tuberculosis	53	22	31	0.397
Mammary cancer	6	3	3 *	0.031 **
Schizophrenia	80	13	67 *	0.77
Down syndrome	89	7	82 **	0.88 *
Spina bifida	72	33	39	0.58
Manic-depression	80	20	60	0.75

A pure-breeding strain of cattle with large horn length (1.5 m) is crossed with a pure-breeding strain with small horn length (10.0 cm), producing an F_1 of intermediate size. These F_1 individuals are interbred to produce a large population (10,000) of F_2 individuals. However from this F_2 generation only 2 had horns as large as the pure-breeding cattle with large horns.

9-3. How many loci would you estimate are involved in horn length in these cattle?

The ratio of cattle with horns as large as the pure-breeding stock in the F_2 generation (2/10,000) would represent the ratio of homozygous (dominant or recessive) individuals. You must therefor determine which polygene system comes closest to that ratio using the formula $1/(2^n)^2$, where n is the number of loci involved.

observed ratio = 2/10,000 or 1×10^{-4}

ratio of homozygous individuals of a;
2 gene system = $1/(2^2)^2$ = 1/16 or 6.25×10^{-2}
3 gene system = $1/(2^3)^2$ = 1/64 or 1.56×10^{-2}
4 gene system = $1/(2^4)^2$ = 1/256 or 3.91×10^{-3}
5 gene system = $1/(2^5)^2$ = 1/1024 or 9.77×10^{-4}
6 gene system = $1/(2^6)^2$ = 1/4096 or 2.44×10^{-4} **
7 gene system = $1/(2^7)^2$ = 1/16384 or 6.10×10^{-5}

A six gene system appears to most closely match the observed ratio.

9-4. Outline the crosses from the parental stock to the F_2 generation using your proposed polygenic model of inheritance.

```
P1        AABBCCDDEEFF      x      aabbccddeeff
F1        AaBbCcDdEeFf      x      AaBbCcDdEeFf
F2   AABBCCDDEEFF, AaBBCCDDEEFF, AABbCCDDEEFF, AABBCcDDEEFF,
     .... aabbccddeeff
```

How many cattle with 10 cm horns would you expect to find in the F_2 generation?

The ratio of homozygous dominant and homozygous recessive individuals would be the same, therefore you should expect approximately 2 of the cattle to have horns of 10 cm.

9-6. Given your polygene model;

a) How much does each effective gene contribute to horn length?
b) What is the horn length of the F_1 cattle?
c) What assumption do you make in determining these values?

a) The absolute difference in horn length is 140 cm, and in a six locus system there are 12 effective alleles. Therefore each gene must contribute 140/12 or 11.67 cm of horn length.

b) The F$_1$ cattle are heterozygous (*AaBbCcDdEeFf*). Starting from the minimal horn length of 10 cm (that of the homozygous recessive), and adding six of the allelic units of length (6 x 11.67), approximately 70 cm, the F2 cattle must have had horns of approximately 80 cm in length.

c) These calculations assume that the loci are equally additive.

9-7. Given that the heritability of egg production is 0.30, what would the mean egg production of the parental stock have to be if a farmer wanted to increase the egg production of his chicken coop from 35 to 40?

Knowing that heritability is calculated as gain/selection differential, use the formula to solve for the mean parental yield.

$$H = \frac{Y_o - \bar{Y}}{Y_p - \bar{Y}}$$

$$0.30 = \frac{40 - 35}{Y_p - 35}$$

$$0.30(Y_p - 35) = 5$$

$$0.30\,Y_p - 10.5 = 5$$

$$0.30\,Y_p = 15.5$$

$$Y_p = 51.67$$

The farmer would need to breed chickens whose mean egg production is 51.67

Due to a severe coastal storm a population of field mice become isolated on a sandy barrier beach. After 250 years of intense selection due to predation, the loci for pelage color become fixed for bleach fur (that is the population is homozygous for bleach fur). Despite the fact that the loci for pelage color are fixed, there is still variation in coat color ($s^2 = 0.13$) within the population. The mainland population from which the island population was separated shows greater variation in coat color ($s^2 = 0.41$).

9-8. Estimate as best as possible the heritability of pelage color for the island population.

Heritability is a measure of the amount of variability of a trait that is due to genetic variability. In this hypothetical case the whole population is homozygous for this trait - there is no genetic variability within the population. Therefore, the heritability of pelage color is zero.

9-9. a) What is the variance due to genetic factors of the mainland population of field mice?

b) What is the heritability of pelage color for the mainland population?

a) The variance in pelage color observed on the mainland represents the total phenotypic variance, $V_{Ph} = 0.41$. Since the genotype for pelage color of the island population is fixed (all are homozygous), then the variance observed there must be due to environmental factors. Therefore $V_E = 0.13$. The variance due to genotype is the difference between these two variances.

$$V_{Ph} = V_G + V_E$$
$$0.41 = V_G = 0.13$$
$$V_G = 0.28$$

b) Heritability (in the broad sense) of pelage color is defined as;

$$H_B = V_G/V_{Ph}$$

Substituting in the observed and calculated variance values produces;
$$H_B = 0.28/0.41 = 0.68$$

9-10. Schizophrenia reportedly has a very high heritability ($H_N = 0.85$). If you were to do a study of cases of schizophrenia to test this heritability, what would you predict the observed correlation among siblings to be?

Using the formula;

$$H_N = r_{obs}/r_{exp}$$

and knowing that r_{exp} for siblings is 0.5, and assuming H_N is 0.85, then;

$$0.85 = r_{obs}/0.5$$
$$r_{obs} = 0.425$$

Key Concepts and Terms

Active site	energy (delta G)
Transformation	Mutability
Prion	*in vitro*
Nucleoside	Nucleotide
Pyrimidine	Purine
Guanine (G)	Adenine (A)
Thymine (T)	Cytosine (C)
Polymerize	Uracil (U)
X-ray crystallography	Phosphodiester bond
Chargaff's rule	Tetranucleotide hypothesis
Complementarity	Double helix
Antiparallel	Polarity
Genetic code	Denature
A DNA	B DNA
Template	Z DNA
Conservative mode of replication	Semiconservative mode of replication
Density-gradient centrifugation	Dispersive mode of replication
Theta structure model of replication	Autoradiography
Replicon	Y-junction
Continuous replication	DNA polymerase (I, II, and III)
Primer	Discontinuous replication
Leading strand	Okazaki fragment
Processivity	Lagging strand
Primase	RNA polymerase
Endonuclease	Exonuclease
DNA ligase	Proofreading
Primosomes	Initiator proteins
Single-strand binding proteins (ssb proteins)	Helicase
Holoenzyme	Replisome
Linkage number (L)	Supercoiling
Topoisomerases (type I and II)	Topoisomer
Rolling-circle model of replication	DNA gyrase
D-loop model of replication	

Study Questions

10-1. Outline the lines of evidence from the 1940's and 50's which lent support to the hypothesis that nucleic acids were the genetic material.

A. The work of Avery, Macleod, and McCarty (1944) showed that the presence of DNA and not proteins, carbohydrates, or lipids was necessary for the transformation of one strain of bacteria (R-type) into another strain (S-type).

B. Hershey and Chase (1952) showed, through the use of radiolabeling (^{32}P for nucleic acids and ^{35}S for proteins), that it is the nucleic acids of bacteriophage (T2) that enter a host cell during the infection process and not proteins.

C. Fraenkel-Conrat and Singer (1956) demonstrated, by reconstituting viral particles using the protein and RNA components from different strains of tobacco mosaic virus, that the characteristics of the resultant virus infection was dependent on the origin of the nucleic acid.

10-2. Outline the lines of evidence which indicated that DNA normally exists as a two-stranded moiety.

A. X-ray crystallography of DNA, primarily of Rosalind Franklin and Maurice Wilkens, was indicative of a helical molecule of two nucleotide strands.

B. Chargaff's ratios of equal amounts of adenine and thymine, and cytosine and guanine suggested the duality and complementary nature of two strands in DNA molecules.

C. Chemical nature of the nucleotides of DNA is such that thermodynamically stable hydrogen bonds form between adenine and thymine, and between cytosine and guanine, thus forming the "rung bridges" between two strands of polynucleotides.

10-3. If the genomic DNA of a particular organism consists of 10% thymine, what is the % base composition of the other three nucleotides?

Based on Chargaff's rule, a given amount of DNA contains adenine and thymine in a ratio of 1:1 and cytosine and guanine in a ratio of 1:1. Therefore, if the DNA consists of 10% thymine, it should also consist of 10% adenine. This leaves the remaining amount of DNA (80%) to consist of equal amounts of cytosine and guanine - 40% each.

10-4. How would the "melting point" of a given amount of DNA that consists of 10% thymine compare to the melting point of the same amount of DNA that consists of 35% guanine?

The temperature at which DNA denatures, or "melts," is determined by the number of hydrogen bonds of the molecule. Since C-G pairing involves 3 hydrogen bonds versus 2 hydrogen bonds between A-T, the more cytosine and guanine in the DNA the higher the melting point. The DNA sample with 10% thymine would be 80% C-G (see question 10-3). The DNA sample with 35% guanine would contain 70% C-G. Therefore the first DNA sample would have a slightly higher melting point than the second.

10-5. During DNA replication how is the single stranded DNA template read by DNA polymerase III, 3'-5' or 5'-3'?

Replication proceeds along the DNA template 3'-5', synthesizing the new DNA strand 5'-3'.

10-6. a) What steps are necessary in discontinuous DNA replication?

b) What enzymes are involved in the steps of discontinuous DNA synthesis?

    a) 1. A DNA-RNA primer fragment must be constructed.
        2. DNA synthesis 5'-3' (elongation).
        3. The removal of the RNA primer.
        4. The replacement of the RNA primer with DNA.
        5. The final ligation of the nascent Okazaki fragment to the preceding fragment.

    b) 1. RNA polymerase (or primase) synthesizes the DNA-RNA primer.
        2. DNA polymerase III is responsible for fragment elongation.
        3. DNA polymerase I acts to excise the RNA primer.
        4. DNA polymerase I also replaces the RNA fragment with the appropriate DNA fragment.
        5. DNA ligase forms the final phosphodiester bond between adjacent Okazaki fragments.

10-7. The following is a 5 base single-stranded DNA fragment.

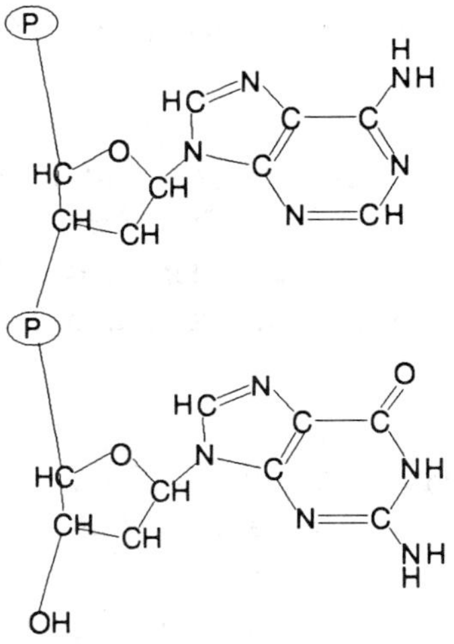

a) Label the 3' and 5' ends.

b) Identify the nucleotides.

c) Draw the appropriate molecular structure for the complementary strand, labeling its' ends and nucleotides.

d) Indicate the hydrogen bonding between the bases.

e) Indicate the proper direction of DNA replication.

10-7 a-e)

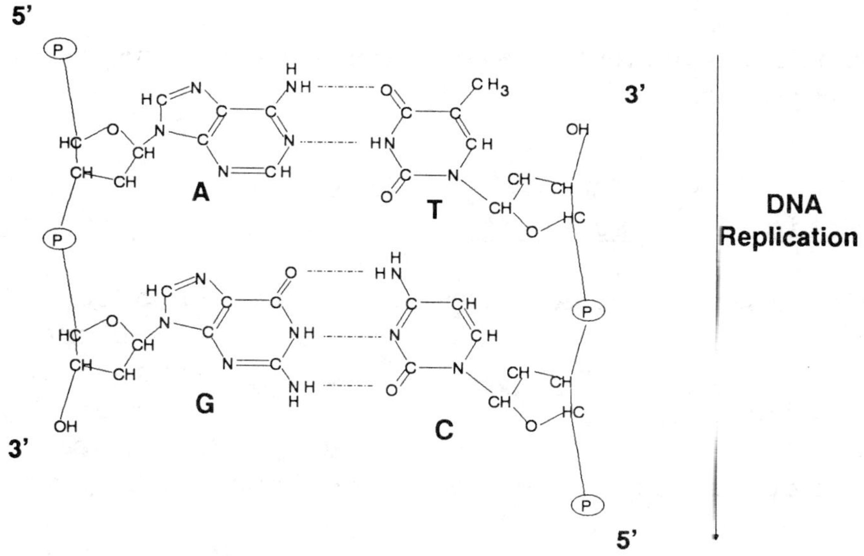

10-8. List the steps for the initiation of DNA replication.

1. The identification of the origin site of replication.

2. The opening of the double helix at the origin of replication.

3. Initiation of the replication at the replication fork for the leading and lagging strands.

10-9. Describe the three models of circular DNA replication and how they differ.

1. The *theta*-structure model allows for replication without breaking the sugar-phosphate backbone of the DNA. Replication is initiated at a particular origin of replication site, and proceeds in two directions (two Y-junctions) and on both DNA strands (continuous and discontinuous).

2. The rolling-circle model of replication proceeds by causing a break in a phosphodiester bond of the sugar-phosphate backbone of one of the DNA strands at a particular site. Replication proceeds along the intact circular strand (discontinuously) and the linear strand (continuously). In this model, DNA replication proceeds in one direction (only one Y-junction).

3. The D-loop model of replication is similar to that of the *theta*-structure except each of the two DNA strands have its own origin of replication site. Replication begins at one site and proceeds along only one strand of the circular chromosome. Only when the origin site of the second strand is exposed does replication proceed along it in the opposite direction. Both strands are replicated continuously.

10-10. List the prokaryotic DNA polymerases, their analogous eukaryotic polymerases and the function they perform.

```
                        DNA Polymerases
Prokaryotic             Eukaryotic analog            Function

DNA polymerase I     DNA polymerase Beta     DNA repair enzyme
                                             short tracts of DNA
                                             (primer replacement)
DNA polymerase II           --               back-up repair enzyme to
                                             polymerase I
DNA polymerase III   DNA polymerase alpha    major replicating enzyme
                                             (continuous replication
                                              and Okazaki fragments)
        --           DNA polymerase gamma    major replication enzyme
                                             of mitochondria
        --           DNA polymerase delta    thought to function
                                             similar to polymerase
                                             alpha
```

Key Concepts and Terms

Transcription
mRNA
rRNA
DNA-RNA hybridization
Promoter
Footprinting
Consensus sequence
Coding strand
Anticoding strand
Heat shock proteins
Rho protein
Rho-dependent terminators
Leader
Svedberg unit, s
Anticodon
Unusual bases
TATA box (Hogness box)
Transcription factors
Primary transcript
Cap
Intervening sequences (introns)
Ribozyme
Group II introns
TACTAAC box
Reverse transcriptase
RNA replicase

Central dogma
Translation
tRNA
Ribosome
Sigma factor
Terminator sequence
Conserved sequence
Pribnow box (-10 sequence)
Template strand
Noncoding strand
Proofreading
Rho-independent terminators
Stem-loop structure
Trailer
Codon
Aminoacyl-tRNA synthetase
Posttranscriptional modifications
CAAT box
Enhancer
Poly-A tail
Heterogeneous nuclear mRNA (hnRNA)
Exons
Group I introns
Splisosome
Small nuclear ribonucleoproteins (snrnps)
RNA phage

Study Questions

11-1. Modify the following diagram of the "central dogma" to include all transfers now known to be possible.

replication

DNA $\longrightarrow$ transcription $\longrightarrow$ RNA $\longrightarrow$ translation $\longrightarrow$ PROTEIN

11-1 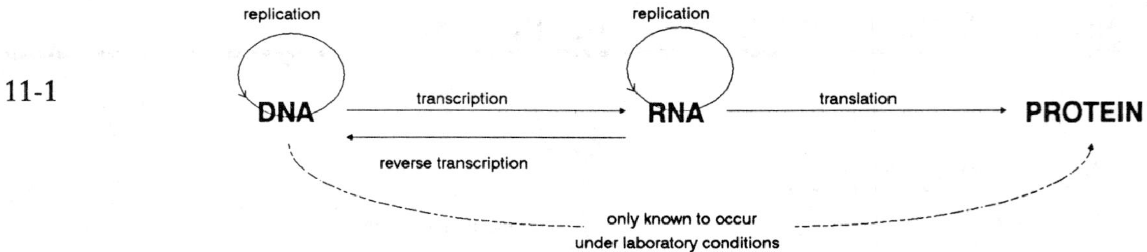

11-2. Describe the different types of RNA. How are they the same? How do they differ?

All RNA is created through transcription of DNA using the bases guanine, cytosine, adenine, and uracil (instead of thymine) and used in the process of translation (protein synthesis) to produce proteins. RNA's differ in their function in the process of translation though. Messenger RNA (mRNA) functions as a template of the specific amino acid sequence of the protein to be produced. Transfer RNA (tRNA) functions to transport the amino acids to the site of translation. Ribosomal RNA (rRNA) functions in the structural integrity of the ribosome and its association with mRNA.

11-3. What are conserved and consensus sequences? What is their significance?

Conserved sequences are nucleotide or amino acid sequences of a gene that are exactly the same among different organisms. Consensus sequences refr to regions of a gene that, although contain variation, are still similar in their base sequence. The conservation, or high degree of similarity, of nucleotide sequences of a gene among different species indicates that that particular domain of the protein is important, and cannot vary much if the protein is to remain functional.

11-4. The following is a segment of DNA (the template strand) with the first base transcribed identified.

Identify the following;

a) the 3' and 5' ends
b) the pribnow box
c) the +15 base
d) the downstream and upstream direction
e) the mRNA produced
f) the initiator codon
g) the nonsense codon
h) the leader sequence
i) the trailer sequence

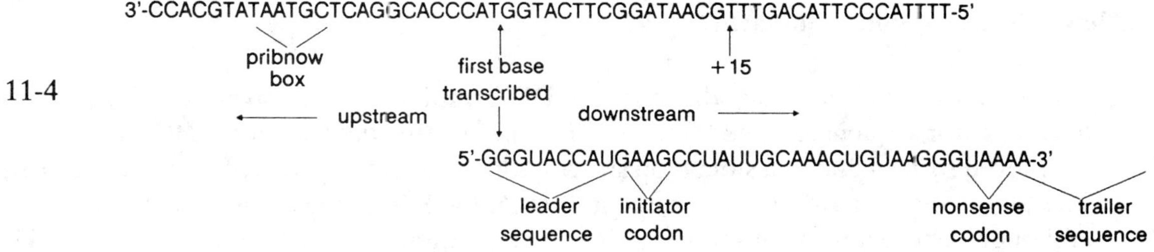

11-4

11-5. Describe the structure of tRNA and how it relates to its function.

All tRNAs have the same basic structure which coincides with its function during translation of associating with the ribosomal-mRNA complex, other tRNAs, specific amino acids, and the alignment of those amino acids for their eventual peptide bonding. tRNAs consist of three stem-loop structures formed by inverted repeats separated by a region which contains unusual bases. The D-loop (the stem-loop proximal to the 5' end) and the T-loop (proximal to the 3' end) function in tRNA-tRNA associations. The middle stem-loop is the anticodon loop involved in associating with a given codon on the mRNA transcript. The 3' end of the tRNA is the site of amino acid attachment under the control of aminoacyl-tRNA synthetases.

11-6. List the different consensus DNA regions that can promote transcription in eukaryotes and their location.

```
TATA box (Hogness box)      -25
CAAT box                    -70
Enhancer region             200
```

11-7. List the posttranscriptional modifications that occur to mRNA in eukaryotes.

1) Poly-A tails added to the 3' end of mRNA.
2) 7-methyl guanosine cap added to the 5' end of mRNA.
3) Introns removed from the mRNA.
4) Exons spliced together.

11-8. What is the nucleolar organizer?

The nucleolar organizer, the site of the nucleolus and ribosomal production, refers to the regions of DNA in eukaryotes which code for multiple copies (130 copies in *Drosophila melanogaster*) of the larger r RNA segments (5.8s, 18s, and 28s). The regions are tandemly positioned on the X and Y chromosomes, and are transcribed by RNA polymerase I as opposed to RNA polymerase II which is the major enzyme of transcription.

11-9. How is transcription terminated?

Transcription continues along the length of DNA until the RNA polymerase reaches a stop, or terminator sequence. If the terminator is a rho-independent sequence, RNA polymerase is believed to slow due to a stem loop structure located after a series of uracils is transcribed. This allows the separation of the nascent mRNA, the RNA polymerase, and the DNA. Although the exact mechanism of termination is unknown, the region of less stable A-U base pairing (with only 2 hydrogen bonds) of the DNA-RNA complex is thought to be involved in the release.

If the terminator sequence is rho-dependent, the presence of the rho protein is necessary for termination otherwise transcription will continue("read-through"). Rho-dependent terminators do not have the A-T rich region on the DNA, but the rho protein, which proceeds along the nascent mRNA 5'-3'. As the RNA polymerase slows at the stem-loop structure it is thought that the rho protein catches up to the polymerase, releasing the DNA-RNA complex by a helicase action of "unwinding" the nucleic acids.

11-10. What are introns and exons, and what is their function?

Introns are intervening sequences of mRNA that are excised before translation. Exons are the coding regions of mRNA that are joined after the excision of introns, and the regions that are translated to produce the functional protein. The function of exons is evident in the production of the specific functional proteins. The exact function of introns is still debated. A currently popular hypothesis is that the presence of introns allows for a diversity of proteins to be produced through recombination ("intron shuffling") of various polypeptide domains coded for by the exons (i.e. globin gene families, LD lipoprotein receptors).

Key Concepts and Terms

Secondary structure
Quaternary structure
Initiation codon
Initiation factors
Scanning hypothesis
P site
Peptidyl transferase
Translocase (EF-G)
Release factors (RF)
Polycistronic
Signal peptide
Docking protein
Polynucleotide phosphorylase
Degenerate code
Mixed families

Primary structure
Tertiary structure
Wobble phenomenon
Initiation complex
Shine-Dalgarno hypothesis
A site
Elongation factors (EF-Ts, EF-Tu)
Translocation
Nonsense codon
Polyribosome (polysome)
Signal hypothesis
Recognition particle
Frameshift
Cell-free system
Unmixed families

Study Questions

12-1. What are paracodons?

Paracodons refer to the characteristic of tRNAs to recognize and bind to specific amino acids. The exact mechanism of recognition of tRNAs to a specific amino acid and subsequent binding by its aminoacyl-tRNA synthetase is unknown.

12-2. Briefly describe the three stages of translation.

<u>Initiation</u> refers to all of the factors involved in the initial setup of the translation process. This is a complex phenomenon involving the formation of a functional ribosome, the association of the initiation codon (5'-AUG-3'), and the N-formylmethionine tRNA ($tRNA_m^{Met}$) in prokaryotes, or unmodified methionine tRNA ($tRNA_1^{Met}$) in the case of eukaryotes.

<u>Elongation</u> refers to the repetitive process of the arrival of the proper aminoacyl-tRNA at the ribosome (based on the association of its anticodon with the current mRNA codon at the A site) and subsequent peptide bond of the amino acid to the preceding amino acid.

Termination of protein synthesis results when a nonsense codon is shifted into the A site of the ribosome, triggering the action of release factors and the dissociation of the ribosomal subunits, the mRNA, and the tRNA.

12-3. What are polysomes? Do they occur in prokaryotes or eukaryotes?

Polysomes are complexes of several ribosomes translating the same piece of mRNA (at different sites). Both prokaryotes and eukaryotes have polysomal complexes, however only prokaryotes can exhibit sequential transcription (mRNA synthesis from DNA) and translation (protein synthesis from mRNA) of a given gene. Eukaryotic mRNA must undergo posttranslational modifications and transport out of the nucleus to the site of protein synthesis before translation can occur.

12-4. How do antibiotics work? Why are fungal infections much harder to treat clinically than bacterial infections?

Many antibiotics kill bacteria by disrupting the cellular process of protein synthesis, either at the level of transcription or translation. Clinically prescribed antibiotics are most useful in the treatment of bacterial infections because they take advantage of the differences in protein synthesis between prokaryotes and eukaryotes. Fungi are eukaryotic, therefore any substance which would disrupt protein synthesis of a fungus would have the same affect on the protein synthesis of the infected person.

12-5. The likely function of polynucleotide phosphorylase in bacteria is to degrade mRNA. Why would a cell degrade its own mRNA after going through the process of transcription to synthesize it?

Different proteins are necessary at varying periods of a cell's life cycle and in varying quantities. The degradation of mRNA under varying conditions and at varying rates of degradation is one way to regulate protein synthesis. If mRNA were not degraded, protein synthesis would continue unchecked, wasting cellular energy and perhaps leading to a toxic buildup of the protein.

12-6. What are mixed families and unmixed families of codons? List the unmixed families and their amino acids.

Mixed family codons refer to codons in which the third base determines the amino acid, or nonsense codon to be coded. Unmixed family codons refer to codons in which the third base is degenerate, that is the first two bases determine the amino acid regardless of the identity of the third base.

Unmixed family codons	Amino acid
UC_	serine
CU_	leucine
CC_	proline
CG_	arginine
AC_	threonine
GU_	valine
GC_	alanine
GG_	glycine

12-7. In which codons is the specific nucleotide of the third base rather than its chemical moiety (that is purine *versus* pyrimidine) important in the decoding process? What do these codons code for?

Although the third base of codons UGA and UGG are both purines, they code differently. UGA is a stop, or nonsense codon, whereas UGG codes for tryptophan. The same is true for AUA and AUG, an isoleucine codon and the "start" methionine codon respectively. In all of the other mixed family codons the chemical moiety of the third base determines the amino acid.

12-8. Below is an mRNA sequence. Give the corresponding amino acid sequence.

3′-AGUAUGCUAUAGGUGUUGUAUUCCGUA-5′

This is partially a trick question because the mRNA strand is presented 3′ to 5′ but should be read 5′ to 3′. The amino acid sequence is;

        Met-Pro-Tyr-Val-Val-Asp-Ile-Val

12-9. For the mRNA sequence of question 12-8, what would the resultant polypeptide be if the 8Th, 14Th, and 16Th bases were deleted?

        Met-Pro-Leu-Leu-Asp-Ile-Val

12-10. Rewrite the mRNA sequence of question 12-8, indicating the bases that could be changed by mutation without any change in the amino acid sequence.

This time writing the mRNA sequence 5′ to 3′, the * indicates those bases that could experience wobble.

        5′-AUGCCUUAUGUUGUGGAUAUCGUAUGA-3′
              *   * *   * *

Key Concepts and Terms

Recombinant DNA technology
Restriction endonucleases
Hybrid plasmid
Hybrid vector
Passenger DNA
Cosmids
Blunt-end ligation
Genomic library
Southern blotting
Western blotting
Expression vectors
Heteroduplex analysis
Transform
Transgenic
Biolistic
Restriction map
Partial digest
Hypervariable loci
Polymerase chain reaction (PCR)
Ladder gels

Gene cloning
Genetic engineering
Palindrome
Hybrid vehicle
Chimeric plasmid
Charon phages
Poly-dA/poly-dT technique
Complementary DNA (cDNA)
Probe
Northern blotting
Dot blotting
Chromosome walking
Heteroduplex mapping
Transfection
Electroporation
Restriction digest
Restriction fragment length polymorphisms (RFLPs)
Double digest
Variable-number-of-tandem-repeats (VNTR) loci
Stepladder gels

Study Questions

13-1. What are restriction endonucleases?

Restriction endonucleases are enzymes that bacteria use to cut foreign DNA and thus protect themselves from invasion, primarily by bacteriophages. The enzymes cleave the DNA at certain sequences. Based on the type of cut made in the DNA and the sequences recognized, restriction endonucleases have been divided into three groups. Type II endonucleases are the ones that are useful to recombinant-DNA studies because they generally recognize sites with twofold symmetry (palindromes) and cut only at those sites. The cuts are usually staggered, resulting in free, single-stranded ends that can form double helices with other DNA treated with the same endonuclease. Thus treatment with these enzymes provides a mechanism to have DNA from different sources joined together.

13-2. What are the problems that can result from inserting foreign DNA into various plasmids?

A fear that many people felt when this technology was developing was that a gene that could cause cancer or some other deadly disease might be inserted into a plasmid or virus of the common intestinal bacterium, *E. coli*. The result could be the unleashing of this disease throughout the human population. To avoid this possibility, scientists developed methods of containment, both biological and physical. Biological containment meant working with bacteria that were severely weakened and plasmids that could not be transmitted between bacteria. Physical containment meant working in facilities of increasing security and containment as the risk increased. It appears now that these procedures are more than adequate to safeguard humanity from an unleashed plague. In fact, even normal bacteria that have been manipulated probably do not pose a risk because they are usually at a competitive disadvantage among the wild-type bacteria. Intestinal bacteria have been living closely with higher mammals for millions of years and have had ample opportunity to spread plagues or pick up toxic genes but have not.

13-3. What is a hybrid vector? How are they selected?

A hybrid vector is a plasmid or phage that has had a piece of foreign DNA inserted. Let's look at an example involving a bacterial plasmid. The simplest hybrid vectors are formed by using a plasmid that is cut once in a nonessential region by a restriction endonuclease. The piece of foreign DNA of interest is isolated by having been cut on either side by the same restriction endonuclease. Then, the cut vector and foreign DNA are mixed; in some instances the appropriate hybridization will take place. That is, a single piece of foreign DNA will be inserted in the cut in the vector. Repair DNA processes can then ligate the DNA. The hybrid vector can then be taken up by bacteria. To select only bacteria that have a hybrid vector, plasmids are used that have two different antibiotic-resistance genes with the cloning cut in the middle of one of these genes. Bacteria that have not taken up any vector will be sensitive to both antibiotics. Bacteria that have taken up a vector with no insert will be resistant to both antibiotics. Bacteria that have taken up a hybrid vector will be resistant to one antibiotic but sensitive to the antibiotic whose resistance gene was cut by the endonuclease; the inserted foreign DNA will interrupt the gene and thus make the bacterium sensitive to that antibiotic.

13-4. What are other methods of cloning besides simple cuts with restriction endonucleases?

DNA of interest can be isolated by methods that do not involve cutting with a specific restriction endonuclease. If the DNA has "blunt ends," that is, no single stranded tail, it can be joined to another piece with blunt ends by a phage enzyme, T4 DNA ligase. These blunt ends would be found if the DNA were isolated by physical shearing or by restriction endonucleases that leave blunt ends. DNA can also be cloned by several methods that create "sticky ends" on the DNA. Alternatively, linkers can be attached to the blunt ends with T4 DNA ligase. Linkers are short pieces of DNA that have a restriction site within them. Thus, once attached, they can be treated with the specific restriction endonuclease and end up with the single-stranded ends characteristic of that restriction endonuclease. Another method used prepares passenger and vehicle DNA for joining by creating complementary ends on each. In the Poly-dA/poly-dT method, a poly-A tail is created on the foreign DNA and a poly-dT tail is created on the vector at its cloning site (or vice versa). The two DNAs can then be joined with hybridization followed by some patch repair to fill in any gaps. The polynucleotide tails are created with the enzyme deoxynucleoside transferase plus a quantity of the triphosphate nucleotide of interest.

13-5. How do we select for a piece of DNA to clone?

In order to clone a segment of a genome, it is necessary to have a double-stranded piece of DNA. If we start with mRNA, which is sometimes easier to get (e.g., hemoglobin mRNA in red blood cells), we can convert it to double-stranded DNA with the enzyme reverse transcriptase. The result is called complementary DNA (cDNA). If RNA is not available, it is sometimes possible to create the RNA or DNA by artificial synthesis if the sequence of the target protein is known. Then the synthesis can be carried out, directed by the genetic code dictionary for the particular amino acids. Alternatively, if those methods are not available, the entire genome of an organism can be broken up into pieces. These pieces, called a genomic digest, can be probed, or, the genomic digest can be cloned, in a random or shotgun fashion, creating a genomic library. From this, a particular gene can be probed for after it has been cloned.

13-6. How is a particular gene located within a genomic library?

A genomic library is a series of clones comprising the entire genome of an organism. These clones can be probed for a specific gene of interest with a radioactive complementary piece of genetic material (single-stranded RNA or DNA), by the process of Southern blotting and hybridization. The probe can be obtained by isolating mRNA if that is available. For example, the RNA for certain genes (e.g., hemoglobin in red cells, tRNA genes, histone genes, rRNA genes) is very easy to isolate. If this is not feasible, then a probe can be created synthetically using the codon dictionary if the amino acid sequence of the protein is known. There are problems in this method, including problems created by the redundancy of the genetic code; a probe consistent with the genetic code dictionary might not be the complementary sequence as the gene. However, depending on how similar the pieces of DNA are, hybridization is possible. It is also possible to locate a particular gene in a genomic library if the gene is expressed. Then, the protein product of that gene can be located by Western blotting. Other methods of locating or developing a probe require ingenuity on the part of the investigator.

What is chromosome walking?

Chromosome walking is a technique for discovering what is located on longer stretches of a chromosome on either side of a known gene. The technique makes use of overlapping probes. To begin with, a gene located by probing is itself converted into several probes by isolation of the gene and breaking it down into several smaller pieces. These smaller pieces are then used to probe a genomic library. The investigator is attempting to locate a cloned piece of DNA in the genomic library that is adjacent to or overlapping the original gene of interest. Since a genomic library is created by randomly breaking chromosomes, it is expected that not only will a gene be cloned once, but adjacent pieces will be cloned with overlapping regions. If one of those overlapping regions is used as a probe, it will locate a cloned piece that overlaps and probably goes further down the chromosome in one direction. By repeating this process of locating an adjacent region and then using that cloned DNA as a probe, it is possible to isolate in clones a long continuous region of a chromosome. These individual clones can be isolated and then either sequenced or examined in other ways to determine what genes, if any, they contain. This technique fails when a region that is repeated in other parts of the genome is located because then, as a probe, it will locate many clones that are not adjacent to the one of interest and thus the walking process will be halted.

13-8. What is heteroduplex analysis?

Heteroduplex analysis is a method to see the length and location of a piece of DNA that has been cloned. In its simplest form, DNA-DNA hybridization is done between a vector without a cloned insert and the same vector with the cloned insert. Since homologous regions will form double helices, there will be a single-stranded loop formed by the insert since it has no complementary homologue. The length of this loop can be measured to verify the length of the insert. In addition, since circular vectors are cut in known places to linearize them for hybridization, the location of the insert can be located in respect to the known ends of the linear vector.

13-9. What vectors are used in eukaryotes?

There are naturally occurring plasmids in yeast. In addition, prokaryotic plasmids have been adapted to yeast by inserting the yeast centromeric region into these plasmids. Thereafter, these plasmids replicate along with the yeast chromosomes during divisional processes. In mammals, the most popular vector used is SV40, a transforming virus. It can be manipulated like phage lambda and can either grow vegetatively in the cell or can be incorporated into a host chromosome. In plants, the crown-gall-tumor system is used extensively. A plasmid of the bacterium, *Agrobacterium tumefaciens*, called the *Ti* (tumor-inducing) plasmid, can be manipulated like many prokaryotic vectors. This plasmid can be used to introduce foreign genes into plants to determine their effects.

13-10. What methods are used to get foreign genes into eukaryotic cells?

The process of integration of foreign DNA into a eukaryotic cell is called transformation; the cell or organism that has been transformed is called transgenic. Animal and plant cells with their cell walls removed can take up DNA directly from the environment. Additionally, the DNA can be injected directly into cells, a technique that has proved successful with mouse cells. Retroviruses can also be used to transfect a cell. If the retrovirus has been modified to include a piece of foreign DNA as well as being made noninfectious, it can be taken up by cells with the aid of normal helper viruses. The modified virus can then integrate into the host chromosome and thereby express its foreign DNA. Foreign DNA can literally be shot into cells by a biolistic process in which the DNA is coated on microprojectiles (e.g., tungsten). DNA can also enter a cell when the cell's membrane is momentarily disrupted by high-voltage electricity. This process is called electroporation. Finally, foreign DNA can enter a cell by incorporating the foreign DNA into artificial membrane-bound vesicles. The process is called liposome-mediated transfer. As time goes on, other methods surely will be developed for the delivery of foreign DNA into host eukaryotic cells.

13-11. What is a restriction map?

A cloned piece of DNA, a vector, or an relatively small piece of DNA can be cut by one or many restriction endonucleases. The product of this process is called a restriction digest. A particular restriction endonuclease can cut a piece of DNA in none, one, or many places, depending on the occurrence of the recognition sites of the enzyme, which, in turn, is dependent on the length of the sequence that the enzyme recognizes; four-cutters recognize many more sites than six-cutters on a piece of DNA. The position of these cuts can be mapped on the piece of DNA, producing what is called a restriction map. Two basic strategies are used: first, map the cuts of restriction endonucleases separately and then map the relative positions of the cuts in relation to each other. The total number of cuts is assessed in a complete digest followed by electrophoresis to determine the sizes of the pieces. The location of the pieces in relation to each other is usually determined by a partial digest in which not all possible cuts are made, resulting in the individual pieces plus larger pieces of DNA that contain uncut segments. By end-labeling the DNA to determine the end pieces,complete and partial digesting followed by electrophoresis shows the location of each piece in relation to its neighbors. After this process is carried out with several restriction endonucleases, the overlap pattern can be determined by doing double digests, treating the DNA with both restriction endonucleases at the same time and then electrophoresing the restriction digest. Like piecing together the pieces of a jig-saw puzzle, the arrangement of pieces of DNA, and hence the location of the cuts, is discovered.

13-12. What are RFLPs?

The term RFLP is an acronym for restriction fragment length polymorphism. This refers to polymorphisms in the bands in a restriction digest indicating that different samples (clones) have changes at sites acted on by restriction endonucleases. For example, if in one clone a restriction endonuclease cuts a particular piece of DNA into 700- and 900-bp fragments, there will be a band at each of these positions on the electrophoretic gel of the digest; there will be no band at 1600 bp. However, if a mutation occurs in another clone such that the site of the endonuclease is changed, the cut will no longer be made. The latter digest will have a band and 1600 bp and none at the two other positions, 700 and 900 bp. This difference is a RFLP and can be used to study mutation, function, and evolutionary history. In addition, these polymorphisms can be used to determine relatedness of individuals; every offspring has some combination of the polymorphisms of its parents. Of particular value are polymorphisms of hypervariable loci, those with many different alleles in a population. VNTR (variable-number-of-tandem-repeats) loci generate enough polymorphisms to be of value in forensic science. For example, the "Jeffreys" probe locates a VNTR locus with enough variation to provide the equivalent of a fingerprint in which each person has a unique combination of bands on a gel of the digest.

13-13. What is PCR?

Polymerase chain reaction (PCR) is a technique to amplify a very small sample of DNA. If the technique is to be used effectively, the sample of DNA must lie between known regions on both strands. The reason for this is that there must be oligonucleotides available (synthesized) that can serve as primers for each strand of the DNA on opposite sides of the DNA of interest. The primers are added, the DNA is denatured, the primers hybridize to their complementary regions, and DNA synthesis occurs on both strands across the central region. This is one cycle of DNA replication, creating two copies of the sample of DNA where only one was present to begin with. In this cycle, the temperature is changed several times: there is a denaturing temperature, an annealing temperature, and a replicating temperature. Normally, since the denaturation temperature destroys the DNA polymerase, the process is stopped, the products purified, and then new components added to repeat the process. You can see that this is very time consuming, and with each purification step, products are lost, leading to low efficiency of replication. However, with the discovery of a DNA polymerase from a thermophilic (heat loving) bacterium, *Thermus aquaticus*, the process can be automated because the denaturation step no longer destroys the polymerase. PCR machines are now readily available that cycle temperatures precisely and quickly. Thus the sample of DNA is added as well as the primers, triphosphate nucleotides, and the polymerase. After about 20 cycles, a million copies or more of the DNA are now present. This technique has had a major impact in certain areas of molecular genetics.

13-14. What is dideoxy sequencing?

There have been several methods developed to sequence segments of DNA. In one, DNA replication takes place in different vessels such that the replication is stopped at one of the bases in each vessel. Rather than stopping the replication at each occurrence of that base in each vessel, only a particular fraction of the time will replication be stopped at a base (e.g., stop replication 10% of the time when a thymine is encountered). After the process has gone to completion, the newly replicated pieces of DNA are electrophoresed, forming bands of known base length representing all the places in which replication was stopped at thymine. For example, if there were a thymine at positions 4, 7, and 10, there will be bands of 4, 7, and 10 bases on the gel. When the products of the four vessels (for the four nucleotides of DNA) are electrophoresed on the same gel, the sequence of the original DNA can be read directly. Stoppage of replication at a particular base is achieved by using a certain fraction of dideoxy bases. For example, we would use 10% dideoxyadenosines in the example above to get termination 10% of the time when a thymine is encountered. Dideoxynucleotides have no OH-group, just a hydrogen at both the 2' and 3' positions of the sugar. Since the primer configuration for continued DNA replication requires a 3'-OH, the dideoxynucleotide causes termination of replication after it is incorporated into the growing DNA chain.

13-15. What are some practical benefits of recombinant-DNA technology?

Recombinant-DNA technology has been of tremendous benefit in medicine, agriculture, and industry for two reasons. First, recombinant-DNA technology has allowed us to understand the way genes work, and second, the technology has allowed us to isolate, refine, or create genes of interest and then amplify them or their products. The latter is especially important to medicine in which substances like growth hormone, interferon, and many others were available in such small quantities at such great cost that effective research and treatment was hampered. Many industries have developed to create and produce these products. Currently there is also an initiative to map and sequence the entire human genome, an expensive and time-consuming project that can benefit us by eventually localizing all of our genes and providing approaches to deal with the diseases that result when these genes don't function. In addition, industry is proceeding in areas of new technology such as developing yeast that can create alcohol from cellulose; the alcohol can then be used as an alternative fuel. In agriculture this technology is being used to develop bacteria that can protect plants from freezing and from insects and perform nitrogen fixation to grow plants better in poor soils. As time proceeds the benefits of recombinant-DNA technology are becoming commonplace.

Key Concepts and Terms

Inducible system	Repressible system
ß-Galactosidase	Lac operon
ß-Galactoside acetyltransferase	ß-Galactoside permease
Repressor	Regulator gene
Operon	Operator
Constitutive mutants	Allosteric protein
Trans-acting	Cis-dominant
Cyclic AMP	Catabolite repression
Corepressor	Catabolite activator protein (CAP)
Attenuator region	Derepressed
Leader peptide gene	Leader transcript
Attenuator stem	Terminator
Site-specific recombination	Preemptor stem
Transposable elements	Antiterminator protein
Polar mutants	Transposons
IS elements	Insertion sequences
Cointegrate	Composite transposon
Heat-shock proteins	Selfish DNA
Codon preference	Antisense RNA
Idling reaction	Stringent response
Relaxed mutant	Stringent factor
N-end rule	Feedback inhibition
PEST hypothesis	

Study Questions

14-1. What is an operon? What are its parts and how does it function?

An operon is a prokaryotic gene arrangement that serves to control the expression of a group of genes that are usually coding for enzymes controlling steps in the same pathway. These genes are transcribed as a single unit with a single promoter; the operon then is a group of genes under the control of an operator, which is a sequence in the promoter of the first gene of the operon. The operator sequence is recognized by a repressor protein that, when bound to the operator, prevents the transcription of the that operon. In those operons that are induced by the presence of a metabolite in the environment, the repressor protein is inactivated by the metabolite (or a byproduct) thus freeing the operon to be transcribed. In those operons that are repressed by excessive metabolite presence, the repressor is not functional until it combines with the metabolite in excess. Repressor proteins are allosteric proteins; they have binding sites for two different substances. However, binding to one of the substances makes the protein incapable of recognizing the other substance, or, alternatively, the protein does not recognize its first substrate until it binds its second substrate that then effectively changes the allosteric protein's shape.

14-2. What are the differences between inducible and repressible operons?

An inducible operon (e.g., *lac* operon of *E. coli*) is one that controls catabolic activities such as the breakdown of a sugar. The enzymes that are coded for in the operon are not needed in the cell until the substrate appears in the environment of the cell. In these cases, the operon normally is repressed. However, when the substrate appears in the environment it binds to the repressor and changes the repressor's shape, causing it to fall free of the operator and thus allow transcription. A repressible operon (e.g., the *his* operon of *Salmonella*) is one that codes for enzymes involved in synthesis, such as the synthesis of amino acids. Normally, these operons are not repressed; the repressor cannot recognize the operator site. However, when excess quantities of the end product of the pathway appear in the cell, they bind with the repressor and convert it into a functioning repressor that then binds at the operator and stops further transcription. Thus, depending on whether the allosteric repressor is active without or with a substrate, the operons can be inducible or repressible.

14-3. What do "*cis*-dominant" and "*trans*-acting" refer to in operon functioning?

There are two units that must interact in order for an operon to function appropriately: the repressor protein and the operator sequence. The repressor protein is coded for by a regulator gene. Either the regulator gene or the operator can mutate to forms that no longer interact appropriately with the other. If the regulator gene mutates, producing a mutant repressor protein, that protein may not bind to the repressor and therefore operon control is lost. This type of mutation works through normal gene action of transcription and translation and a mutation of the regulator gene will affect any copies of its target operon that happen to be in the cell at the time; we call this "*trans*-acting" control. However, a mutation of the operator region that results in an operator that no longer appropriately recognizes its repressor will affect only that operon of which the operator is a part; we call this a "*cis*-dominant" effect. "*Cis*" means on the near side and "*trans*" means across.

14-4. What is a constitutive mutation of an operon?

By the basic mechanism of operons, they must be repressed (turned off) at appropriate times. That is, inducible operons should be repressed when no substrate is in the environment, and repressible operons should be repressed when too much end product of their pathways is present. A constitutive mutation is one that results in an operon never being repressed; the operon is transcribed all the time. This type of effect can come about by mutations of either the regulator gene that codes for the repressor or the operator sequence. Mutations of either that prevent their recognizing the other will result in a repressor protein that will never bind to its operator. We call this constitutive transcription and the mutation is a constitutive mutation.

14-5. What is catabolite repression?

When given several sugars that they are capable of metabolizing, many prokaryotes will metabolize glucose in preference to the others. In other words, with several sugars appearing in the environment, each capable of inducing a different operon, only glucose is metabolized; the other operons do not seem to be induced. This phenomenon is referred to as catabolite repression. In fact, the other operons are induced; however, their level of transcription still remains very low in the presence of glucose. The mechanism involves cyclic AMP. Operons that are capable of undergoing catabolite repression have CAP sites in their promoters. When bound with a complex of a catabolite activator protein (CAP) and cyclic AMP (cAMP) the transcription rates of these promoters are enhanced. Without either CAP or cAMP, transcription rates, even in induced operons, are very low. It seems that the presence of glucose in the cell reduces the quantity of cAMP by either inhibiting adenylcyclase, the final enzyme in the pathway of cAMP production, or stimulating phosphodiesterase, the enzyme that converts cAMP to AMP. Without cAMP the various sugar operons with CAP sites are inhibited by the failure to have their transcription enhanced.

14-6. What is attenuator control of an operon?

Some repressible operons are also controlled by a second mechanism, attenuator control, in which transcription of the operon is dependent on the ability of a small gene within the promoter region to be translated. In some amino-acid coding operons this mechanism works in addition to normal operator repression; in some operons it is the only mechanism of transcriptional control. In attenuator-controlled systems, a leader-peptide gene exists between the promoter and the first structural gene of the operon. This leader-peptide gene has codons for the amino acid being synthesized by the proteins coded for by the operon. Thus in order for the leader-peptide gene to be translated, there must be tRNAs charged with the appropriate amino acids. Depending on whether translation proceeds or not, different secondary structures form in the mRNA being transcribed from the operon. The RNA of this gene is called the leader transcript. One configuration of the leader transcript RNA causes the cessation of transcription of the operon; another configuration allows transcription to proceed. Thus the translation of this gene tests the occurrence of the amino-acid product of the gene; if there are adequate quantities then translation of the leader-peptide gene will proceed but transcription of the operon will cease. If there is an inadequate supply of the charged tRNA, then translation will stop but transcription of the operon will continue.

14-7. What are the lytic and lysogenic cycles of phage lambda?

Many bacteriophages can pursue either of two different life-cycle phases after infecting a bacterium. In one life-cycle phase, the phage can replicate and lyse the cell; this is called they lytic cycle. In the other, the phage can integrate into the bacterial chromosome and be replicated along with the bacterial chromosome. The phage will not lyse the cells. This life-cycle phase is called lysogeny. Phages that can enter lysogeny are called temperate phages; the host bacteria are called lysogenic. Phage lambda is a temperate phage. It can enter either life-cycle phase, depending on a competition between two repressor proteins in the phage chromosome.

14-8. What controls whether a temperate phage integrates into its host or enters the lytic cycle?

The phage lambda chromosome is composed of four operons. The repressor operon contains two antagonistic repressor genes, *cI* and *cro*. These two repressors compete for operator sites for the left and right operons. If *cI* is successful then lysogeny will follow because the operons that code for all of the lytic function of the phage are repressed but the phage can still integrate into the host chromosome. If the *cro* repressor outcompetes the *cI*

repressor, then the lytic cycle will follow because the other operons will be transcribed and translated whereas the *cI* gene will be inhibited. Hence the "choice" of which life-cycle alternative to follow is determined by the competition to these repressors, a competition determined to some extent by factors within the environment of the infected bacterial cell.

14-9. What are transposons?

Transposon is a shortening of the term transposable genetic element; it refers to a section of DNA that is capable of moving a copy of itself to a new place in the genome of the organism. The transposon minimally contains genes for the transposition process. It is proximally surrounded by an inverted repeat and then distally by a direct repeat on the host chromosome. The direct repeat comes about as an outcome of the transposition process; the inverted repeat is probably a signal to the transposing enzymes of the limits of the transposon itself. Simple transposons are also called insertion sequences or IS elements. Complex transposons can form by having a central region of DNA surrounded by two IS elements. Virtually any DNA between two IS elements can be transposed, deleted, or inverted. The target site is not a specific sequence on the host chromosome.

14-10. What is the mechanism of transposition?

Shapiro suggested a mechanism of transposition involving a fusion or cointegrate state. The transposon is inserted into its new location through staggered cuts, which generate the direct repeat seen at the boundary of transposons. The model requires the action of the two genes carried by functioning transposons, transposase and resolvase. In essence, the target site is cut and the transposon is joined to the target site while still attached at its original position. One strand of the transposon double helix is separated from the other and through the process of patch repair using DNA polymerase I, two copies of the transposon are generated. Then, the cointegrate state is resolved through crossover to the original situation with the exception of the fact that a copy of the transposon has been inserted in a new location.

14-11. How are bacterial flagellar types controlled by transposition?

In the case of *Salmonella*, transposition can affect the phenotype by determining the type of flagellar protein produced. Flagellar protein can be of the phase-1 or phase-2 types controlled by the *H1* or *H2* genes, respectively, depending on which gene is transcribed and translated. Gene *H2* is in an operon with a repressor for gene *H1*. Thus if that operon is transcribed, the phase-2 protein will be produced and the *H1* gene will be repressed. If the *H2* operon is not transcribed, then only the *H1* gene will be transcribed. The promoter for the *H2* operon is in a transposon that is capable of inversion. In one state, the operon is transcribed leading to phase-2 protein. In the inverted state, the *H2* operon cannot be transcribed and the bacterium has the phase-1 protein.

14-12. What are heat-shock proteins and how are they regulated?

Most proteins in the bacterium *E. coli* are transcribed by RNA polymerase with the sigma^{70} factor. A group of 17 genes in *E. coli* are transcribed with the aid of a different sigma factor, sigma^{32}. Under heat stress, these 17 proteins appear and apparently help the cell withstand the effects of elevated temperature, a response seen in virtually all organisms. In order for the heat-shock proteins to be transcribed, the sigma^{32} must appear in the cell, which is the first stage in the heat-shock response. In other words, elevated temperatures leads to the activation of the *HtpR* gene, the gene for the sigma^{32} protein. It then follows that the heat-shock proteins are transcribed.

14-13. What is antisense RNA and how can it control gene expression?

Once a gene is transcribed, there are still mechanisms to prevent expression of the gene. One way for that to happen is for the mRNA to be bound to a complementary RNA and form a double-helical structure, specifically at the 5' end of the RNA, so that it would not be bound to a ribosome for translation. The RNA complementary to a transcribed gene is called antisense RNA. There are several cases known in which gene expression is controlled in this way. For example, in *E. coli*, the *ompF* gene, a gene that codes a porin protein is controlled by an antisense RNA. Although the likely source of an antisense RNA would be the complementary strand of the DNA, it appears in this case that the antisense RNA comes from another porin gene, *ompC*. Why this control exists is not well understood, but the actual mechanism seems to be clear.

14-14. What is the stringent response of a cell?

When a cell is being starved, it is to that cell's advantage to shut down most physiological process until such time as the crisis is past. A cell can sense starvation by the existence of uncharged tRNAs in the A site of ribosomes. When a cell senses this situation at the ribosome, an odd nucleotide, 3'-ppGpp-5' (guanosine tetraphosphate) is produced. This production is part of what is called an idling response--the ribosome is idling without a charged tRNA. This odd nucleotide is produced by an enzyme known as the stringent factor and the nucleotide leads to what is called the stringent response, which includes a major cessation of transcription.

14-15. What is feedback inhibition?

The last place in which control of gene expression can be exerted is at the point in which an enzyme, whose gene has been transcribed and then translated, is prevented from functioning. One way for this to happen is if the enzyme is an allosteric protein, one with two active sites such that binding of a substrate at one site prevents the activity of the second site. In this way, many enzyme pathways are stopped by allosteric binding of the end product of the pathway with one of the first enzymes in the pathway. For example, the enzyme threonine dehydratase is the first enzyme in the pathway of converting threonine to isoleucine. The second site on the enzyme binds isoleucine; the allosteric enzyme then cannot catalyze the transformation of threonine to alpha-ketobutyrate. Thus, excessive quantities of the end product of the pathway, isoleucine, can stop continued activity of the pathway. We call this feedback inhibition.

Key Concepts and Terms

Nucleosomes	Uninemic
Nuclease hypersensitive sites	Histone
Nonhistone proteins	Scaffold
Balbiani rings	Chromosome puffs
Satellite DNA	Giemsa stain
Constitutive heterochromatin	Euchromatin
Telomeres	Intercalary heterochromatin
DNA-DNA hybridization	Telomerase
Unique DNA	Cot values
Gene family	Repetitive DNA
Gene amplification	Alu family

Study Questions

15-1. What are the major differences between eukaryotes and prokaryotes?

The essence of prokaryotes is simplicity; the essence of eukaryotes is complexity. Prokaryotes are small, with little internal structure, no nucleus, and are primarily single-celled organisms. Eukaryotic cells are larger with membrane-bound internal structures, including the nucleus. Although some eukaryotes exist as single-celled organisms, most develop into multicellular organisms with differentiated tissues. Prokaryotes have small circular chromosomes; most transcription is controlled by operons; transcription is coupled to translation; and introns are missing. Eukaryotes have linear chromosomes, transcription and translation occur at different places (nucleus vs. cytoplasm), and introns in genes are the rule.

15-2. In eukaryotes, does each chromosome consist of a single, linear DNA molecule?

The answer appears to be yes, eukaryotes do have chromosomes that consist of a single, linear DNA molecule. This information comes from several sources. First, radiolabeling studies have confirmed the uninemic nature of the eukaryotic chromatid. Chromosomes were allowed to replicate in tritiated thymidine and then placed in cold medium. Based on semiconservative replication of uninemic chromatids, we expect, after one round of replication in unlabeled medium, that for every chromosome entering mitosis, one chromatid will be labeled and one unlabeled. This prediction has been validated. In addition, predictions about further distribution of the label after more generations of growth in unlabeled media have all been verified. Additionally, it has been verified that the sizes of the DNA molecules in an organism are of the size predicted based on a uninemic structure when viscoelastic studies are done or when large DNA molecules are isolated carefully. Thus all experimental work is consistent with the view that the eukaryotic chromosome is uninemic.

15-3. What is the composition of eukaryotic nucleoprotein?

When the chromatin of a eukaryotic organism is analyzed, it is found to consist of DNA, proteins, and RNA in a weight ratio of 100:150:1, respectively. The RNA is most likely transcriptional products bound to the chromosomes and not a significant structural part of the chromosomes. The protein can be divided into histone and nonhistone proteins in a ratio of about 3.5:1.0, respectively. The histones are a group of five homogenous proteins that are invovled in the general organization of the chromatin, not with control of gene expression. The nonhistone proteins are made up of scaffold proteins that provide the general shape of the chromosomes as well as any regulatory proteins that may be involved with control of gene expression.

15-4. How can we isolate large numbers of a particular chromosome for study?

Individual chromosomes can be isolated by the technique of high-speed flow cytometry, in which individual chromosomes are recognized and then separated from others in a mixture of chromosomes from disrupted cells. Chromosomes are identified with ultraviolet lasers and photomultipliers based on their fluorescence after being treated with two dyes, Hoechst 33258 and chromomycin A3. These dyes fluoresce at different wavelengths and each chromosome binds different quantities of the two dyes. Thus treated chromosomes can be recognized by their differences in fluorescence. In flow cytometry, chromosomes are identified and then the liquid they are in is vibrated to form minute droplets. The droplets with the chromosome of interest are charged and then separated from the rest with charged plates. In this way, chromosomes are identified and then isolated. The rate of isolation is about 200 chromosomes per second (0.1 ug/hour of DNA) at about 90-95% purity.

15-5. What are nucleosomes?

In mitosis and meiosis, eukaryotic chromosomes contract down to very compact structures in order to divide successfully. The packaging of the DNA is accomplished by coiling. The first order of packaging is to have the DNA associate with protein in a nonspecific way. The proteins are the histones and the association of two coils of DNA with one histone structure is referred to as a nucleosome. There are five fractions of histones, H1, H2A, H2B, H3, and H4. Two molecules each of histones H2A, H2B, H3, and H4 form the core of the nucleosome. Histone H1 appears to be involved in the entrance and exit of the DNA in the nucleosome; only one copy of histone H1 is involved for every two copies of the others. About 146 bases of DNA are associated with the nucleosome with another 50-75 base pairs, called a linker, between nucleosomes. The coiling of nucleosomed DNA leads to the higher-order structure of the chromatin during the replication processes of mitosis and meiosis.

15-6. What are nuclease hypersensitive sites?

Most DNA in eukaryotes is intimately associated with histone proteins as nucleosomes. However, some regions do not appear associated with histones. These regions were discovered by treating the DNA with nucleases. Only regions not associated with histones can be attacked by the nucleases because only histone-free DNA is not protected by the nucleosomes. These regions are called nuclease hypersensitive sites. When these regions are looked at in detail they are found to contain sequences indicative of active sites involved in replication or transcription of the DNA. Thus most DNA is routinely in association with histones except areas that normally are available for enzymes that initiate processes of replication or transcription. These areas are kept free of histones in ways that we are not yet aware of.

15-7. What are polytene and lampbrush chromosomes?

Aside from eukaryotic chromosomes in their normal appearance that we see during mitosis and meiosis, chromosomes in certain organisms, at certain times, in certain tissues, take on different morphologies. Two examples are the polytene and lampbrush chromosomes. Polytene chromosomes are seen in the salivary glands and other tissues of some diptera and are formed by endomitosis in which the chromosomes form about 1,000 copies by DNA replication, but the copies synapse with each other rather than separating. The result is a large "cable" of about 1,000 copies of the chromosome in close association. Under the microscope we see distinct bands that are invisible in single copies of the chromosomes. These chromosomes have been of major importance in the development of cytogenetic concepts.

Lampbrush chromosomes are found in amphibian oocytes in which very active transcription takes place. Remember that these are the giant frog's eggs found in ponds and are full of nutrients for the developing tadpole. The chromosomes are in a looped-out configuration, undergoing active transcription. They reminded early cytogeneticists of lamp brushes, items unfamiliar to most students today.

15-8. What is the significance of chromosome puffs and Balbiani rings?

Chromosome puffs are diffuse areas seen in polytene chromosomes. Balbiani rings are puffs; the term was originally coined for puffs in the midge but now is used synonymously for the largest of the puffs. Chromosome puffs are areas of active transcription. The many copies of the DNA in the polytene chromosome are looped out, presumably for ready access by the transcribing enzymes and free nucleotides. During development, chromosome puffs occur under several different kinds of conditions, consistent with the needs of the cell for certain gene products. Puffs can be stage specific, tissue specific, environmentally induced, or constitutive.

15-9. What is the significance of chromosome banding?

Under various types of treatments, different banding patterns can be made to appear in eukaryotic chromosomes. These bands are consistent and have been proven to be useful for two very different reasons. First, they allow us to differentiate chromosomes within the same cell that are very similar in size and morphology. For example, Down syndrome was found to be trisomy 21 after the G-banding pattern of human chromosomes was worked out. Second, the bands tell us something about the composition of the DNA. Treatment with Giemsa stain brings out G-bands, which are really bands already faintly visible in the mitotic chromosomes as groups of small chromomeres (darker regions of the chromatin). Presumably the G-bands make up intercalary heterochromatin, chromatin not normally transcribed. A variant of the Giemsa technique brings out another type of bands, C-bands, for centromeric bands. These bands, as the name implies, are found in the centromere and are presumably made of centromeric heterochromatin, a form of heterochromatin giving structure to the centromeric regions. This type of DNA forms satellite bands during ultracentrifugation. Finally, the regions between the G-bands can be stained; these bands are referred to as R-bands, or reverse bands. Presumably they represent the transcribable, euchromatic regions of the chromosomes.

15-10. What is a centromere?

Centromeres are constrictions in eukaryotic chromosomes where microtubules bind to kinetochores during mitosis and meiosis. Using genetic-engineering techniques, it has been possible to isolate the DNA of yeast centromeres. They all are about 200 base pairs (bp) long with three consensus regions. The length of the entire region is about 150-200 ., just about the diameter of a yeast microtubule, indicating that only one microtubule attaches at each yeast centromere. However, the yeast model is a simple one and may not represent the details of higher eukaryotes. Specifically, most higher eukaryotes have more than one microtubule attach at each centromere. In addition, yeast centromeres seem to lack both the obvious constriction and the centromeric heterochromatin. Hence, we expect higher eukaryotic centromeres to be more complex than yeast centromeres.

15-11. What is a telomere?

A telomere is the end of a linear chromosome, a feature of which every eukaryotic chromosome has two. A telomere must mark the end of a chromosome such that it isn't degraded by exonucleases nor is it treated as a broken or "sticky" end by enzymes that might try to attach chromosome ends, yielding fused chromosomes. Sequence analysis has shown that telomeres are repeats of five- to eight-base sequences, repeated 250-1,000 times. In human beings, the telomeric sequence is TTAGGG. This sequence is found in all vertebrates and even in some trypanosomes, although other sequences appear in ciliates and yeast. The enzyme telomerase adds telomere sequences to the ends of chromosomes in a template-free manner. The number of copies of this sequence is variable, but kept within limits by some mechanism that has not yet been discovered. There also seems to be some unusual hairpin configurations of the DNA in the telomeres, configurations whose functions are unknown.

15-12. What is a cot curve? What does it indicate?

The term cot refers to units of time and concentration of single-stranded DNA. Under circumstances of uniform concentrations of DNA, the cot value is a time value for the process of renaturation of denatured DNA. If DNA is broken into small pieces and heated, it will form single-stranded DNA. If that DNA is then cooled, it will begin to renature: regions of complementary sequences will come in contact and double-stranded regions will reform. The rate of that renaturation is dependent on the complexity of the sequence of DNA in the sample, all other things being equal. For example, DNA that is repetitive of the base cytosine on one strand and guanine on the other will renature almost instantly (snap-back renaturation). DNA that has no repetitions of any sequences will renature the slowest, and DNA with some repetitive sequences will renature at an intermediate time. By experimentation with nucleotide samples of known sequences, it was possible to translate the time (cot) value into a length of unique sequence of the DNA. Thus, a new sample of DNA could be treated similarly to form a cot curve and from the midpoint of the curve it is possible to know the length of unique DNA. Eukaryotes produce cot curves indicating three regions: unique, intermediately repetitive, and highly repetitive DNA at 70, 15, and 10%, respectively.

15-13. What is the significance of the types of DNA found in eukaryotic genomes?

From the analysis of cot curves, it was found that eukaryotes have unique, intermediately repetitive, and highly repetitive DNA at 70, 15, and 10%, respectively. The highly repetitive DNA is made of about 1,000,000 repeats of a sequence about 200-bp long. This is the satellite or centromeric heterochromatin fraction of the eukaryotic genome. Clearly, this DNA is serving a structural purpose around the centromeres. Intermediately repetitive DNA is made of several components. First is dispersed DNA, such as the Alu family, that appears throughout the genome in many copies. Second are numerous copies of genes of known function that are needed in many copies because of high transcription rates, such as ribosomal RNA genes and histone genes. Finally, there are genes that appear in many copies but the copies have diverged to some extent, such as the globin family of genes. Unique DNA makes up the bulk of the DNA and presumably represents most transcribed DNA. Hence, beginning with an analysis of cot curves, we can see that the DNA of eukaryotes is a complex mix of at least three different kinds of genetic material.

15-14. What is gene amplification?

Gene amplification is a process in which the cell makes extra copies of genetic material that is needed in very large quantities, above the amount found normally in the genome. An example is the ribosomal RNA genes in some amphibians that are amplified during oogenesis. During this process, oogenesis, the African clawed toad accumulates about 10^{12} ribosomes. Thus a very large quantity of rRNA is needed. The cell provides much of the DNA for these genes by amplification in which small circles of rRNA genes are accumulated. There have been several models put forward to explain the way in which this DNA is created, but the mechanism is not known for certain. It could be due to unequal crossing over or unscheduled DNA replication followed by recombinational events.

15-15. What is the globin gene family?

The term gene family is used for a series of genes that has arisen by duplication to form many copies. In some gene families these copies keep the same sequence and function whereas in others, multiple copies have led to divergence. That is, from an evolutionary perspective, multiple copies of the same gene have given the cell the opportunity to keep the original function and yet copies were available to evolve new gene functions. An example of this is seen in the globin gene family. Globins are oxygen transporting and storage molecules (myoglobin and hemoglobin). In human beings, the myoglobin and hemoglobin genes diverged evolutionarily from the ancestral, primitive globin gene. The ancestral hemoglobin gene, after duplication, diverged into an alpha and beta gene, each of which has diverged into from four to five different functioning genes as well as several nonfunctioning genes, pieces of DNA that remain but are not transcribed or translated. These simply build up mutational differences because they are not under any constraints since they are not expressed. The differences in sequences of all of these genes have allowed evolutionary biologist to be able to trace the evolutionary history of this gene family.

Chapter 16 - Gene Expression: Control in Eukaryotes

Key Concepts and Terms

Fate maps	Development
Homeo box	Homeotic mutants
Helix-turn-helix motif	Homeo domain
Leucine zipper	Zinc finger
Totipotent	Copper fist
Homothallic	Isoschizomers,
Cassette mechanism	Heterothallic
Immunity	Pheromones
Antibodies	Antigen
Cytotoxic T lymphocyte	Immunoglobulins (Ig)
Major histocompatibility complex	T-cell receptors
Idiotypic variation	Allotypes
Hybridomas	Monoclonal antibody
Junctional diversity	V-J joining
N segments	Allelic exclusion
Cancer	Somatic hypermutation
Metastasis	Tumors (neoplasms)
Lymphomas	Leukemias
Carcinomas	Sarcomas
Ataxia-telangiectasia	Xeroderma pigmentosum
Oncogenes	Fanconi's anemia
Anti-oncogene	Cancer-family syndromes
Wilm's tumor	Retinoblastoma
Retrotransposons (retroposons)	Molecular imprinting
Proto-oncogenes	Insertion mutagenesis
	Fragile site

Study Questions

16-1. What properties of the nematode, *Caenorhabditis elegans*, make it ideal for developmental studies? What is an AB.araapapaav cell?

The nematode, *Caenorhabditis elegans*, is ideal for developmental studies because of several properties. First, it has very few cells, about 1,000 in the adult. Therefore it was possible to trace the fate of every individual cell. A system of describing development of cells has been worked out in which each cell is labeled by its position after a division. First, the original cell divides to form major progenitor cells, of which there are six altogether; one is the AB cell which is a progenitor of skin, neurons, and muscles. Then cells are noted as follows: a, anterior; p, posterior; r, right; l, left; d, dorsal; and v, ventral, explaining the nature of the AB.araapapaav cell in the question. *Caenorhabditis elegans* is also an excellent study organism because it grows well in the lab, has a short generation time, and genetic studies have revealed a wealth of genetic diversity.

6-2. What are homeotic mutants? What is a homeo box? A homeo domain?

Homeotic mutants are those in which one cell type follows the developmental pathway of another cell type. For example, the *Nasobemia* mutation of *Drosophila* causes legs to grow where the antenna should be. These mutations are important because they are probably master-switch genes in development; they are genes that control the way in which cells develop and are thus a key to understanding the mechanisms of development. In homeotic genes, there seems to be a common 180-base pair segment; this segment is similar (evolutionarily conservative) in organisms as diverse as yeast and human beings. The homeo domain has the ability to bind DNA, further leading to speculation that homeotic genes are regulatory genes that control the functioning of many genes in development. Study of these genes is a major effort in modern developmental genetics.

16-3. What are some of the known motifs of DNA-binding proteins?

The homeo domain is a DNA-binding polypeptide that has a helix-turn-helix motif. In this design, one helical part of the protein binds to the DNA and one stabilizes the bound configuration, with a short sequence of amino acids between the two (the turn). Other motifs are known to bind to DNA, including the leucine zipper, zinc finger, and copper fist. The leucine zipper consists of a helical region of DNA in which leucines line up on one face. Two of these polypeptide regions can line up, in zipper fashion, forming a structure that binds DNA. The zinc finger is a motif in which a finger-like projection of amino acids forms around a base of cysteines and histidines that bind a zinc ion. The copper fist is a protein formation around copper ions with positively charged "knuckles" that bind DNA. Computers can now search for any of these motifs when a new DNA sequence is uncovered. The presence of one of these motifs is evidence that part of the function of the protein is to bind to DNA and therefore the protein might be regulatory in function.

16-4. What is a totipotent nucleus?

A totipotent nucleus is one that can give rise to all the cell types in an adult organism. That a nucleus in a differentiated cell can be totipotent can be discovered by either transplanting the nucleus from a differentiated cell into a zygote that has had its nucleus removed, or taking a single differentiated cell and having it develop into an adult organism. These methods have been used to show totipotency in amphibians and carrots, respectively, among other organisms. The importance of these findings is that the process of development doesn't permanently change the genetic material of an organism. Therefore development is a process that probably proceeds by the nonpermanent induction and repression of genes. We have a fine understanding of many bacterial and phages systems that work just that way and lead us to believe that we will soon understand development in higher organisms with the same degree of clarity.

16-5. How might methylation and Z DNA be involved in eukaryotic development?

Z DNA was first believed to be an oddity in DNA structure. However, when it was discovered that it could be stabilized in vivo with methylation, it was treated more seriously as a form of DNA that could have an important role in control of eukaryotic transcription. When searched for, Z DNA was found dispersed throughout eukaryotic genomes. The involvement of methylation is significant because in some organisms methylation is related to gene activity. That is, certain genes when not being transcribed are more methylated than when they are being transcribed. This led to the idea that Z DNA might occur in promoters of genes and might prevent transcription of those genes. The signal for transcription might involve the reduction in methylation, which would cause the Z DNA to revert to normal (B) DNA and transcription would then follow. There has been somewhat of a loss of enthusiasm for this ideas since it has not proven to be consistent across all genes and all organisms studied. However, Z DNA and methylation could still prove to be important in control of development.

16-6. What are some systems in which transposons control development in eukaryotes?

Transposable genetic elements can control development in two general ways. In the first, a transposable element can enter a gene and disrupt it, appearing in the phenotype as a maladaptive interruption of a homeostatic system. Alternatively, an evolved system of development could be tied to transposition. Both of these kinds of systems are known. For the former (disruptive cases) we mention controlling elements in maize in which McClintock first discovered transposons. For the latter (evolved cases) we mention the cassette mechanism of yeast in which the organisms frequently switch mating types by transposition of silent (unexpressed) mating-type genes to the active site of expression, the MAT locus.

16-7. What components of the immune system have very high levels of diversity?

High levels of diversity are needed in a functioning immune system to cope with the diversity of foreign molecules invading a healthy organism. To handle the hundreds of thousands or millions of different types of invading molecules, immune systems have developed the ability to generate the same high levels of variation. These appear in three distinct places. Generally, high levels of diversity are found in: antibodies produced by B cells; T-cell receptors of the cytotoxic T lymphocytes; and major histocompatibility complex (MHC) gene products. B cells produce antibodies that circulate in plasma and lymph and coat antigens. All cells have MHC proteins on their surface, partly to indicate "self" to other components of the immune system. When a cell is infected by a foreign invader, such as a virus, the cell presents part of the foreign molecule at its surface in conjunction with MHC proteins. This combination is recognized by the T-cell receptors on the cytotoxic T cells that will destroy the infected cell.

16-8. What are the components of an antibody?

Antibodies, or immunoglobulins, are large proteins produced by B cells. They are composed of two identical heavy chains (approximately 440 amino acids each) and two identical light chains (approximately 214 amino acids each) all held together with sulfhydryl bonds to form a Y-shaped molecule. Each branch of the Y is composed of a light and heavy chain end and the base of the Y is made up of the carboxy terminuses of the heavy chains. The amino ends, at the branches of the Y are composed of variable regions that actually recognized the antigen. The rest of the molecules are made of constant regions. The light chains are both either kappa or lambda molecules and the heavy chains are both alpha, gamma, delta, epsilon, or mu molecules; each B cell produces only one type of light chain and one type of heavy chain.

16-9. How is antibody diversity generated?

An antibody molecule is synthesized by the joining of various segments in the region of that gene by processes of site-specific recombination. We will take as an example, the kappa region of mice. There is not one kappa gene, but instead about 300 variable regions, 5 joining regions, and a constant region. You can see that 306 gene segments can be responsible for 1500 different final molecules, depending on which variable region is attached to which joining region (300 x 5). Further diversity is achieved at the time of joining by variability at the crossover point. In addition, heavy chains have another set of regions called diversity regions. There are theoretically 18 billion different antibodies that can be produced by a single mouse or human being. In addition, there are at least two other mechanisms that seem to add even more diversity, somatic hypermutation and N-segment formation (template-free base additions).

16-10. What are cancer-family syndromes?

Cancer-family syndrome refers to the inheritance of the predisposition to contract cancer rather than the inheritance of a specific type of cancer. Pedigrees of families with these syndromes have many affected individuals although the types of cancer in different members of the pedigree differ. For example, breast, colon, prostate, and pancreatic cancer are found in a cancer-family syndrome pedigree in high frequency. The existence of cancer-family syndromes is used as evidence that some families are predisposed toward cancer whereas most families are not. The underlying mechanism might be the inheritance of one mutation in a two-mutation mechanism for cancer or the inheritance of some genetic damage or predisposition towards cancer.

16-11. What is an oncogene?

An oncogene is a gene that causes cancer. First found in retroviruses of chickens (Rous sarcoma virus), oncogenes are found in viruses and in normal cells as well. They can be activated to cause cancerous growth by several mechanisms: they can be mutated, they can be translocated to a very active promoter, or they can be amplified. Normal oncogene function falls into several categories. These include tyrosine kinases, growth factors, GTP-binding proteins, and other functions. Of the functions known, in each case excessive activity is consistent with cancerous growth.

16-12. What are anti-oncogenes?

Since oncogenes cause cancer, anti-oncogenes act to repress cancer. Anti-oncogenes show up in homozygous recessive individuals that contract a cancer that the dominant gene protects against. Two examples are retinoblastoma and Wilm's tumor. In both cases, cancer can be correlated with the absence (deletion) of a chromosomal region. However, the exact limits of the deletion do not seem important, leading to the view that elimination of all or part of the locus is important, not the exact bounds of the deletion as would be important if some sort of enhanced control were operating. It has also been shown that a normal chromosome inserted into cancer cells growing in culture will cause the cells to revert to the normal condition, another bit of evidence that the disease is caused by an anti-oncogene.

16-13. What genes do retroviruses carry normally?

All functioning retroviruses carry at least three genes: the *gag, pol,* and *env* genes. The *gag* gene codes for core virion proteins. The *pol* gene codes for several proteins including reverse transcriptase, to convert the viral RNA genome into DNA; a protease; and an integrase. The *env* gene codes for envelope glycoproteins. Hence with only these three genes, a retrovirus can code for its own proteins, enzymes needed for replication of its genetic material, and enzymes for alternative life-cycle stages (i.e., integration into the host chromosome). Retroviruses usually cause cancer when they pick up into their genetic material one of the oncogenes. Usually retrovirus oncogenes are identical to cellular genes. However, since they usually lack introns, it is assumed that they were picked up as mRNAs by the retroviral RNA chromosome.

16-14. What treatments, promising for the AIDS virus, are arising out of an understanding of the molecular genetics of retroviruses?

AIDS is not a cancerous disease, but is a disease of the immune system caused by a retrovirus that attacks T4 cells. The AIDS virus receptor is a protein called CD4. With destruction of these cells, individuals become prone to diseases found commonly in people taking immunosuppressive drugs. These diseases, such as pneumonia and Kaposi's sarcoma, have become known as the AIDS syndrome and are fatal. There is currently no known cure for the disease. However, an understanding that the disease is caused by a retrovirus that is bound by the CD4 protein of T4 cells has led to some inroads in containing the disease. First, AZT (3'-azido-2',3'-dideoxythymidine) has been found to be helpful. This thymidine analogue is of value because it is preferentially used by reverse transcriptase whereas the normal DNA polymerases do not prefer it. AZT is a dideoxy nucleotide meaning that it will result in chain termination and thus will selectively kill AIDS virus particles. Another promising treatment is the administration of free CD4 receptor proteins to tie up AIDS viruses without allowing them into cells.

16-15. What is a fragile site?

Fragile sites, as the name implies, are chromosomal sites that are prone to breakage. They are of special genetic interest because several cancers are associated with fragile sites. It seems that when a fragile site breaks there is a possibility that a translocation will occur. Certain translocations result in the induction of cancer by bringing oncogenes adjacent to very active promoters and thus amplifying the transcription of these genes. For example, the cancer, Burkitt's lymphoma, is associated with a fragile site on chromosome 8. The cellular oncogene *c-myc* occurs at the fragile site. Translocation that leads to Burkitt's lymphoma places *c-myc* at the promoter of the immunoglobulin gene, *Ig-Cu*. This promoter is extremely active in B cells, the cells that are transformed in the lymphoma. Hence a fragile site results in translocation of an oncogene to an active promoter leading to the cancer we call Burkitt's lymphoma.

Chapter 17 - DNA: Its Mutation, Repair, and Recombination

Key Concepts and Terms

Fluctuation test
Functional alleles
Cistron
Pseudoalleles
Recon
Intra-allelic complementation
Reversion
Back mutation
Nutritional-requirement mutants
Permissive temperature
Tautomeric shift
Transversion
Suppressor gene
Missense mutations
Antimutator mutations
Photoreactivation
DNA glycosylases
Mismatch repair
SOS response
Homologous recombination
Branch migration
Hybrid DNA
Heteroduplex DNA
Gene conversion

Mutation
Complementation
Cis-trans complementation test
Structural alleles
Muton
Complementation groups
Mutation rate
Point mutations
Intragenic suppression
Temperature-sensitive mutants
Restrictive temperature
Transition mutation
Intergenic suppression
Nonsense mutations
Mutator
Dimerization
Excision repair
AP endonucleases
Postreplicative repair
SOS box
Breakage and reunion
Chi sites
Heterozygous DNA

Study Questions

17-1. Assume that in the process of doing the fluctuation test, Luria and Delbruck measured the following number of colonies from ten "individual cultures: " 9, 8, 8, 8, 8, 8, 9, 9, 9, and 7. What conclusions would they have drawn?

In this case, there is uniformity among the individual cultures in the occurrence of T1-resistant cells. In other words, all ten small cultures had about the same number of cells present that were resistant to phage T1. The conclusion that Luria and Delbruck would have been forced to draw is that there is a low, uniform rate of induction of wild-type cells (sensitive to the phage) to the state of resistance. This could only come about if the "genetic" system of the bacteria were not "normal;" this situation would have prevented the use of bacteria in genetic studies.

17-2. Two recessive wing mutants of *Drosophila* are independently isolated in a stock culture. They each produce a different change to the wing venation. How would you determine if they are functional and structural alleles of each other?

To determine allelism, we would perform a complementation test. The two mutants would be mated to each other and their offspring observed. If the two mutations were of different loci (nonallelic) then they would complement each other and the heterozygous offspring would be of the wild type, with completely normal wing veins. If however, the mutations were of the same locus, then they would not complement each other and the offspring would have some mutant form of the wing veins. These mutations would thus be functional alleles. Next, to determine if the mutations were structural alleles, we would look for recombination in the locus within the heterozygote. There would appear rare wild-type offspring and, in similar numbers, offspring with both mutant sites within the same locus. These two would occur as the result of the rare crossover between the two mutant points if the mutations were not structural alleles. To look for these recombinant offspring, we would testcross heterozygous females with males of either mutant type and look for wild-type offspring. If these occurred (above the background mutation rate) we would conclude that the mutations are within the same locus but do not encompass the same bases. If no wild-type offspring appeared, we would conclude that the mutations were at the same base or bases within the same locus; that is, they were structural alleles.

17-3. Why was the system that Benzer used to study genetic fine structure unsuitable for determining the colinearity of DNA and protein?

Benzer attempted to discover the number of sites within a gene that could recombine or mutate. To do this he developed the system using rII mutants of phage T4. This system allowed him to screen billions of phage in a very short time period; thus the screening system was a complete success for the purposes that it was set up for. However, to determine whether changes in DNA were colinear with changes in protein it is necessary to have both the DNA and its protein product to work with. To determine fine structure (mutation and recombination sites) only required assessing phenotypes of phage on different strains of bacterial hosts; no protein product was needed. Unfortunately, Benzer did not know the protein product of the rII locus in phage T4 and thus could not correlate the position of DNA changes with the position of amino acid changes in the protein product of the gene. That problem was solved by Yanofsky working with the tryptophan synthetase A protein.

17-4. What are the differences between inter- and intra-allelic complementation?

Interallelic complementation is the occurrence of the wild type in a heterozygote; each mutant complements the other, allowing the wild-type phenotype to appear in the offspring of the two mutants. Thus complementation is the standard test for allelism. If two unknown mutants result in wild-type offspring when mated, they complement each other and thus the mutations of each are not allelic. However, under certain conditions, mutations that are allelic can complement if the phenotype is controlled by a protein made of subunits. Under rare conditions, complementation can take place at the polypeptide level in which two different polypeptides, controlled by alleles of the same locus, can produce the wild-type phenotype when they interact. This intra-allelic complementation is thus complementation by allelic mutations, which should normally not complement; the existence of this situation needs to be kept in mind when interpreting the results of complementation tests.

17-5. What is a mutation rate?

Mutation occurs in all genetic material in viruses, prokaryotes, and eukaryotes. We measure its occurrence with a rate, usually a number of mutations per locus per some unit tied to the life cycle of the organim or virus. In multicellular eukaryotes it is the number of mutations per locus per gamete. In prokaryotes and single-celled eukaryotes it is the number of mutations per locus per cell division. In viruses and phages it is the number of mutations per locus per progeny phage particle. Depending on the size of the locus, the environment, and the length of time, mutation rates vary from about 10^{-5} to 10^{-10}.

17-6. What is a point mutation? What are its consequences?

A point mutation is generally meant to mean the change in a single base of the genetic material. Point mutations can consist of replacements, insertions, or deletions. We focus on what these changes do to codons for amino acids and stop information during translation. An insertion or deletion has the effect of changing the frame of reference of reading the genetic code (frameshift mutation). The result of this type of change is usually catastrophic for the protein because all codons during the translational process, from the point of the mutation on, are read in a different frame of reference. New stop codons will appear and codons will not be for the amino acids originally coded for. From the point of the mutation on, the protein will either be shortened or composed of amino acids unrelated to the original code.

Changed codons can be classified as the result of missense or nonsense mutations. A missense mutation changes the codon for one amino acid into the codon for another amino acid. A nonsense mutation converts the codon for an amino acid into one for a stop codon, prematurely stopping the translation of that protein. A nonsense mutation will almost certainly have catastrophic results to the cell whereas a missense mutation can have effects that range from virtually unnoticeable to catastrophic depending on whether the locus can be translated into a protein that catalyzes the appropriate reaction.

17-7. What is a conditional-lethal mutation?

A conditional-lethal mutation is one that results in a phenotype that is lethal under one set of environmental circumstances but not another. For example, a mutation that disrupts a metabolic pathway for the synthesis of an amino acid normally would be lethal to a bacterial cell. However, if the medium in which the cell is growing contains that amino acid, then the cell will survive and compete well with the wild-type cell. Another type of conditional-lethal mutation is the temperature-sensitive type in which a mutated protein functions normally under one temperature (permissive temperature) but is inactive under another temperature (restrictive temperature). Temperature-sensitive mutations have been invaluable in studying cellular functions that are absolutely vital, such as DNA replication. Normally the cell could not tolerate most mutations to DNA polymerases and other enzymes and thus they could not be studied with standard mutational analyses. However, temperature-sensitive mutations can be studied because they are normal under one set of temperature but mutations can be recognized when tested at restrictive temperatures.

17-8. What are possible mechanisms for spontaneous mutagenesis?

Mutagenesis can be caused by chemicals, ionizing radiation, temperature shocks, and other environmental insults. However, mutation can occur spontaneously, independently of all of the above. The general cause is tautomeric shifts in which adenine and cytosine shift from the amino to the imino forms and guanine and thymine shift from the keto to the enol forms. In the tautomerized forms, new base pairing will result during DNA replication. These new base pairings cause transitional mutations, changes of one purine-pyrimidine base pair to another in the same orientation. Transversions can be caused be a combination of tautomeric shifts and base rotations about their glycosidic (base-sugar) bonds. For example, an adenine-adenine base pair can form during DNA replication if one of the adenines is in the imino form and the other is rotated about its glycosidic bond from the anti to the syn configuration. Hence, after another round of DNA replication, an adenine-thymine base pair is converted to a thymine-adenine base pair, a transversion. Additionally, insertions and deletions can take place as a result of misalignment of template and progeny strands of DNA. These are all spontaneous changes in the DNA.

17-9. How do various chemicals cause mutations?

Chemicals cause mutations several ways. In one, involving base analogues, a different base is incorporated into DNA and this base has different pairing rules or tautomerizes at a very high rate. An example is 5-bromouracil, which is incorporated as thymine; however, it tautomerizes to a much greater extent than thymine normally does. Other chemicals change nucleotides to ones with different pairing properties. For example, nitrous acid converts amino groups to keto groups, converting, as an illustration, cytosine to uracil. Since uracil pairs with adenine rather than thymine, a transition results. Other chemicals, such as alkylating agents, remove the bases from the nucleotide residues within the DNA, leaving various options of pairing during DNA replication if these errors are not immediately repaired. Finally, some chemicals can insert themselves into the DNA double helix and cause insertions and deletions during DNA replication. Acridine dyes can do this, presumably because of their flat ring structure that allows them to fit into the DNA double helix.

17-10. What are inter- and intragenic suppression?

Partial restoration of the wild-type phenotype can take place in a mutant cell or organism without the original base-pair change being restored. Second changes, either within the same locus or in other loci, can result in a partial return of the wild type. These changes are referred to as suppression. Intragenic repression can occur as a second insertion or deletion within the same locus that then restores the reading frame of an original frameshift mutation. For example, if a base was inserted into a locus, the result would be a frameshift in the translational reading frame of codons. However, if a second base, in the vicinity of the insertion, were deleted, the reading frame would be restored in all but the region between the two changes. If that region is small, and the cell can survive with the changed amino acids in that region, then an approximation to the wild-type phenotype will be restored. Intergenic suppression usually refers to mutations at tRNA loci that recognize the original errors as something different and partially restore the wild-type phenotype. For example, assume that a nonsense mutation occurs in a gene, causing premature termination of the protein during translation. If a mutation occurred in which a tRNA had its anticodon changed such that it now recognized the termination codon as one for an amino acid, transcription would occur. However, this circumstance could only occur if there were another, similar or identical, tRNA that still had the original anticodon. Hence, these mutations are of genetic interest but not really of evolutionary interest because the organisms having these mutations would not compete well enough to survive in the wild.

17-11. What is site-directed mutagenesis?

Site-directed mutagenesis is a genetic-engineering technique to change selectively the sequence of a gene and thus selectively change amino acids in a protein to study protein function. The method is conceptually very simple. A gene of interest is cloned into a vector and the single-stranded form of that vector is isolated. Then, an oligonucleotide is synthesized that is complementary to a part of the gene and will act as a primer for DNA synthesis. However, that oligonucleotide can be synthesized with one or more bases changed. Under laboratory conditions, the oligonucleotide will still form a double helix with its complementary region and DNA synthesis will occur. The original strand will have the original sequence; new strand will have the mutated sequence. The new strand can be isolated and used as a template to make double stranded DNA. The end result is a cloned gene with a new sequence, made to order for the geneticist to study.

17-12. What are the major mechanisms that a cell uses to repair DNA damage?

Damage repair of DNA falls into three general categories: damage reversal, excision repair, and postreplicative repair. The former two are mechanisms used before DNA with the damage is replicated; the latter is used after the cell has tried to replicate the damaged DNA. Damage reversal is a relatively simple process of finding damaged DNA and restoring the original without any repair replication. For example, thymine dimers can be undimerized by the product of the phr gene in bacteria. Excision repair is a process in which changed, damaged, or missing bases are discovered, excised, and repaired with a small patch of new DNA synthesized with DNA polymerase I. Excision repair is initiated by enzymes that recognize mismatches, missing bases (AP sites), thymine dimers, uracils, and other damage. Under certain circumstances (e.g., thymine dimers), DNA replication cannot take place and the process skips the damaged region leaving a single-stranded segment. In postreplicative repair, part of the sister DNA helix is used to patch the single-stranded region; repair patching is then done to restore two complete double helices. Although the original lesion still remains, the cell has another cell cycle in which to repair it by other means.

17-13. What is the SOS response of cells?

Normally, many of the DNA repair enzymes are partially repressed by a repressor known as LexA that binds at a region in the promoters of these genes known as the SOS box. Under conditions requiring postreplicative repair, single-stranded DNA is formed at sites where replication could not take place. This single-stranded DNA is bound by a protein called RecA, which becomes activated and is involved in the breakdown of the LexA repressor. Thus active transcription takes place at the repair genes with SOS boxes in their promoters.

The RecA protein is instrumental in the postreplicative repair processes, allowing single-stranded DNA to insinuate itself into complementary double-stranded DNA, a process critical to postreplicative repair. After postreplicative repair takes place, there will be no single-stranded DNA to be activated, LexA will bind at the SOS boxes, and the cell will no longer be in the SOS response.

17-14. What are the features of the Holliday model of recombination?

Recombination is a precise mechanism in which two double helices break and unite. Normally, the process is precise to the base pair; in other words, it normally does not generate insertions or deletions of bases. Robin Holliday suggested a model that involved a hybrid DNA molecule as an intermediate form. Breakage occurs in one strand each of two homologous DNA molecules and one strand of each insinuates itself into the other. After a process of branch migration, a second cut of the DNA occurs in each molecule. If the second cuts are in the same strands as the original cuts, the result is called a patch; if the cuts are in the uncut strands, the result is a true crossover, or a splice. In a splice, recombination occurs for loci outside the region of the branch migration. In both processes, splices and patches, heterozygous DNA is formed in the region of branch migration. If there are any mismatches in the region they will either be repaired by excision repair mechanisms or separate without repair at the next DNA replication cycle.

17-15. How can DNA repair account for gene conversion?

During recombination, along the stretch of branch migration, heterozygous DNA is formed; one strand comes from one parent helix, one from the other parent helix. Along this stretch there can be mismatches where the two parental DNA double helices had different base sequences. Mismatch repair enzymes usually use methylation cues to determine which of the strands is new and which is old DNA so that repair processes can remove the new base, assuming that it is the "wrong" one in a mismatch. However, at recombination, both parental double helices usually are fully methylated. Therefore the repair enzyme has no information as to what is new and what is old DNA; repair is therefore random at these mismatches. In yeast for example, assume two alleles, *a* and *b*, differing by one base pair. If, during meiosis in a tetrad, that region falls into the region of branch migration, then two of the double helices will be untouched (one having the *a* and one having the *b* allele). The other two double helices are heterozygous. They can be repaired either of three ways: *a* and *b*, *a* and *a*, or *b* and *b*. Either of the latter two results in a spore ratio of 3:1 (*a* to *b* or *b* to *a*). This is gene conversion, caused by mismatch repair in a heterozygous region after recombination.

Chapter 18 - Extrachromosomal Inheritance

Key Concepts and Terms

Cytoplasmic inheritance
Spiral cleavage
Petites
Dendrogram
Plastids
Binary fission
Autogamy
Kappa particles
Mate killer
Metagon
R plasmids
Transfer operon
Resistance transfer factor

Maternal effects
Dauermodification
Mitochondrion
Buoyant density
Chloroplast
Proplastids
Conjugation
Exconjugants
Paramecin
Mu particles
Sex-ratio phenotype
Col plasmids

Study Questions

18-1. What are the general categories of extrachromosomal inheritance and what are their mechanisms of control?

Aside from nuclear genes (the endogenote in prokaryotes) and the environment, the phenotype of an organism can be controlled by maternal effects and cytoplasmic inheritance. Maternal effects are the result of the influence of the maternal cytoplasm in the development of eukaryotic zygotes. Since most eukaryotic organisms begin development within maternal cytoplasm, the female parent has the opportunity to influence development of the zygote in a different way than the male parent. We see examples of this in snail coiling patterns and pigmentation patterns in moths. Cytoplasmic inheritance refers to inheritance influenced by organelles, particles, and parasitic organisms that have their own genetic material. Hence the inheritance of cytoplasmic traits follows the passage of cytoplasm rather than the inheritance of nuclear or endogenotic genes. Examples of cytoplasmic inheritance are the petites of yeast and antibiotic resistance in *Chlamydomonas*.

18-2. How do we determine if an aspect of the phenotype is controlled by extrachromosomal inheritance?

Extrachromosomal inheritance can be identified by changes in phenotypic appearance or phenotypic ratios over time or aberrations in offspring ratios in reciprocal crosses. That is, when a trait is suspected of being of extrachromosomal origin, it must be scrutinized with care. If the nature of the inheritance pattern cannot be readily determined, than some sorts of manipulations are frequently used. In eukaryotes, nuclei and cytoplasm can be separated from each other with the procedures of nuclear transplantation or heterokaryon tests. Various chromosomes can be separated from their original cytoplasm by repeated backcrosses with the male parental type: female chromosomal genes will be halved each generation but the cytoplasm will not be diluted. In addition, cytoplasmic particles can be looked for and isolated (bacteria or viruses) and various treatments can be used that affect organelle genomes rather than host chromosomes.

18-3. What are the possible genotypes of a sinistral snail? Of a dextral snail?

Snail coiling is ultimately controlled by a chromosomal gene with two alleles; dextral coiling (*D*) is dominant to sinistral coiling (*d*). However, this form of inheritance has a maternal component: the coiling of a snail is determined by its mother's genotype, not its own. Thus a sinistral snail must have had a homozygous recessive mother (*dd*). In that case it must have inherited a recessive allele from its mother and can thus be either homozygous recessive or heterozygous (*dd* or *Dd*), depending on its fathers genotype. A dextrally coiled snail must have had a mother with at least one dominant allele (*D*). Since the mother could have been either homozygous dominant or heterozygous (*DD* or *Dd*), and since the father could have been of any genotype, the dextrally coiled snail could have any genotype, including homozygous recessive, *dd*. The table below shows all possible matings with the genotype and direction of coiling of the offspring snails within the body of the table.

Father's genotype

		DD	Dd	dd
	DD	*DD* dextral	*DD* *Dd* dextral	*Dd* dextral
Mother's genotype	**Dd**	*DD* *Dd* dextral	*DD* *Dd* *dd* dextral	*DD* *Dd* dextral
	dd	*Dd* sinistral	*Dd* *dd* sinistral	*dd* sinistral

18-4. What genes are present in the human mitochondrial DNA (mtDNA)? What genes must be present in order for a mitochondrion to function?

The human mtDNA has been sequenced: it is a circle of 16,569 base pairs (bp). It has genes for 22 tRNAs, 16S and 12S ribosomal RNAs, and 13 polypeptides involved in the formation of enzymes of oxidative phosphorylation. The ribosomal RNAs are used to form ribosomes within the mitochondrion; these ribosomes have prokaryotic affinities, indicating prokaryotic origins evolutionarily. When mtDNAs from various organisms were sequenced it was found that there did not seem to be any general rules for what proteins must be coded by the mitochondrion itself, nor what size the mtDNA must be. It has been suggested therefore that whatever coding is done by the mtDNA, it is a relic of the original prokaryotic invader.

18-5. What determines which proteins synthesized within the cell cytoplasm are transported into the mitochondrion?

Proteins that are targeted for transport into and through membranes have signal sequences, sequences of amino acids at the amino-terminal end of the nascent proteins. In the case of proteins that are destined to enter mitochondria, signal sequences do not seem to have consensus amino acid sequences but seem to have other attributes. They are up to 85 amino acids long, have alternations of positively and negatively charged amino acids, and form alpha helices with hydrophobic and hydrophilic faces. Not only are these attributes found on proteins that are transported into mitochondria, but when the DNA sequences for these signal regions are attached to genes for proteins that normally do not enter mitochondria, they then are transported into mitochondria.

18-6. A haploid petite yeast cell is crossed with a wild-type cell. The diploid offspring undergoes meiosis and the four products are isolated and grown up as separate colonies. All four are wild type. What is the mode of inheritance of the particular petite in this cross?

Petite phenotypes have been found to fall into three categories based on crosses similar to those in the question. Segregational petites result in a ratio of wild type to petite of 2:2, indicating that the mutation is in a chromosomal gene. In the second category, the offspring are usually all wild type, indicating a mutation of a mitochondrial gene. The mutant and wild-type mitochondria are mixed in the diploid cell; all the offspring get both types of mitochondria and the wild type controls the phenotype. Presumably the wild-type mitochondria out compete the mutant forms and eventually supplant them. These petites are called neutral petites. In the third type of petite, also a mutation of the mitochondrial DNA, the offspring are usually petites. Here the mutant mitochondria control the phenotype, perhaps by converting the wild-type petites to mutant form. This type of petite is called a suppressive petite. By the pattern of inheritance shown in the question, the petite seems to be of the neutral type.

18-7. How can we use mitochondrial DNA (mtDNA) to help us know about the evolutionary origins of our own species?

Mitochondrial DNA is especially valuable in evolutionary studies of human beings (and many other species) because of two attributes. First, mtDNA is inherited maternally; therefore, any mutations that occur are passed on directly because there is generally no opportunity for recombination with paternal mitochondria, which do not enter the zygote. Second, virtually all the mitochondria within a single individual are of exactly the same genotype. From these assumptions it is possible to use mtDNA to trace lineages. Mitochondrial DNA can be cut with restriction endonucleases and RFLPs (restriction fragment length polymorphisms) determined. In 1987 Cann et al. did this for 147 individuals from five geographic regions; they found 133 different patterns. They then fed these data into a computer program that creates dendrograms (branching diagrams) of relatedness based on mutational differences among the RFLPs of individuals. The program can create many dendrograms; a minimal-length tree was used that had the minimal number of mutations that could create the data set. There were several consequences of this dendrogram: it defined an original split involving Africa and it indicated that all but the original African population was heterogeneous. The last step was to put a time scale on the process; the mutational clock of the mtDNA was calibrated against known human migrations with the result that the original split was put at about 200,000 years ago. The conclusion is that human beings had origins in Africa at that time and that all present-day people can trace their ancestry to a single woman, known affectionately as "Eve."

18-8. What similarities do chloroplast and mitochondrial DNAs have?

Chloroplasts, like mitochondria, are eukaryotic cellular organelles that we believe evolved from prokaryotic invaders. Chloroplasts, the chlorophyll-containing organelle of plants, are the site of photosynthesis and starch grain formation. Chloroplast-containing cells also contain mitochondria. Both organelles have circles of DNA and ribosomes. The chloroplast has a much larger circle of DNA than the mitochondrion, minimally five times larger. Like the mitochondrion, the chloroplast DNA (cpDNA) has genes for tRNAs, ribosomal RNAs, and some polypeptides found within the organelle; many proteins, however, are imported into the chloroplast.

18-9. Flowers on a corn plant that is green are fertilized with pollen from a corn plant that is variegated. The F_1 plants are all green. If they are self pollinated, what will the F_2 be and in what ratios? What will the reciprocal cross yield?

In this problem we assume that the green parent is of the wild type and the variegated plant is homozygous for the iojap allele (*ij*). Fertilization will result in an F_1 generation that is heterozygous (*Ijij*) and green. Self pollination results in a 1:2:1 ratio of *IjIj:Ijij:ijij*, respectively. The *ijij* homozygotes will be variegated. If the reciprocal cross is done, all genotypes will be the same through the three generations as just outlined; however, the phenotypes will be different. The F_1, of *Ijij* genotype, will be variegated because the female parent was variegated and the F_1 gets its chloroplasts from its female parent--pollen does not contribute chloroplasts to the zygote. Similarly, the F_2 will consist of all three genotypes, yet all will show some form of variegation because of the passing on of apparently defective chloroplasts. These results are due to the fact that the iojap allele induces variegation by interfering with the maturation of chloroplasts. However, once the chloroplast is interfered with, it apparently stays in the proplastid state even if in a heterozygote. Thus variegated plastids are inherited maternally giving differences in the reciprocal crosses outlined above.

18-10. What are the differences between autogamy and conjugation in *Paramecium*?

As the names themselves imply, autogamy is a process involving only one *Paramecium* and conjugation involves two. Both processes have similarities and differences in the nuclear rearrangements that take place. Autogamy is triggered in a single cell; conjugation is triggered when two cells of opposite mating types come together. In both processes, the macronucleus breaks down and the micronuclei undergo meiosis. Seven of the eight nuclei break down, leaving one that then undergoes mitosis to form two identical, haploid nuclei. In autogamy, these nuclei fuse forming a single diploid nucleus. Following two mitoses, two of the four nuclei become macronuclei and two remain micronuclei. At the next division the macronuclei separate while the micronuclei divide. The end result is two cells, each with one macronucleus and two micronuclei as before. The genetic consequence is that the micronuclei are now homozygous at every locus. In conjugation, at the point in which each of the conjugating cells has two haploid nuclei, one each migrates into the other conjugating cell. The nuclei then fuse as the cells separate; the remaining processes are similar to those during autogamy. The genetic consequence here is that both cells (exconjugants) are identical although not necessarily homozygous.

18-11. How do we know that the killer trait in *Paramecium* is caused by a cytoplasmic element?

The key piece of evidence that the killer trait of *Paramecium* was cytoplasmicin origin was the observation that a killer cell could make its conjugating partner a killer only under circumstances in which cytoplasm was exchanged. Light and electron microscopic examination led to the finding of a bacteria-like particle, named *Caedobacter taeniospiralis*, that is found in killer cells but not in sensitives. Although there are complications in the life cycle of the bacteria (to become a killer cell, the resident bacteria must be attacked by a bacteriophage), its presence is inexorably linked with the killer trait, and thus its cause.

18-12. How do we know that alleles at two nuclear genes are needed in order to maintain the mate-killer trait in *Paramecium*?

Paramecium is an organism in which genetic studies are relatively easy to carry out. In order to show that alleles at two loci are needed to maintain mate killer, the appropriate crosses need to be carried out. For example, there are homozygous stocks that cannot maintain the mate-killer trait. Since they have arisen through autogamy, they are known to be homozygous. In some cases, crosses between these stocks yield stocks that can maintain mate killer. The implication is that alleles of at least two loci are needed, each stock contributing one of them. If exconjugants are allowed to undergo autogamy, we predict that only one fourth will maintain the mate-killer trait, a prediction borne out. The sequence of crosses is shown; M_1 and M_2 are the alleles at the two loci needed for maintenance of mate killer and m_1 and m_2 are alleles that will not maintain the trait.

$$M_1 M_1 m_2 m_2 \quad \text{X} \quad m_1 m_1 M_2 M_2 \qquad \text{(neither stock maintains mate-killer trait)}$$

$$\downarrow \text{ conjugation}$$

$$M_1 m_1 M_2 m_2 \qquad \text{(mate-killer trait maintained)}$$

$$\downarrow \text{ autogamy}$$

$$M_1 M_1 M_2 M_2 \qquad \text{(genotype that maintains mate-killer; 1 in 4)}$$

$$M_1 M_1 m_2 m_2 \qquad \text{(genotype that does not maintain mate-killer; 1 in 4)}$$

$$m_1 m_1 M_2 M_2 \qquad \text{(genotype that does not maintain mate-killer; 1 in 4)}$$

$$m_1 m_1 m_2 m_2 \qquad \text{(genotype that does not maintain mate-killer; 1 in 4)}$$

All crosses made (those shown here and others) are consistent with the need for dominant alleles at two loci to maintain the mate-killer trait.

18-13. How do we demonstrate that sex-ratio trait in *Drosophila* is extrachromosomal in origin?

The sex-ratio trait in *Drosophila* is one in which females produce mostly, if not exclusively, female offspring. There are both chromosomal and extrachromosomal forms. Extrachromosomal inheritance can be demonstrated several ways. First, by genetic crosses with marker chromosomes, it is possible to take a stock of sex-ratio flies and substitute every chromosome by ones from a stock of flies without the sex-ratio phenotype. However, even with every gene replaced, the trait is maintained. Second, cytological examination shows that about half of a sex-ratio female's eggs do not develop (presumably the male eggs). Cytoplasm can be withdrawn from these undeveloped eggs and used to infect other females with the trait, demonstrating that something in the cytoplasm is the infective agent. In fact, that infective agent has been isolated and is a spirochete.

18-14. What are Col and R plasmids?

Bacteria harbor a variety of plasmids in their cytoplasm. You are familiar with the F factor, which is a plasmid. In addition there are plasmids that carry resistance to various antibiotics; these are referred to as R plasmids. For example, plasmid R222 carries resistance to streptomycin, sulfonamide, chloramphenicol, and tetracycline, originally in the *Shigella* bacterium responsible for a form of dysentery. Then, there are plasmids that carry genes for proteins, called colicins, that are toxic to other strains of bacteria. There are many types of colicins; some enter the attacked cell, many do not. Plasmids can exist in a nontransferrable state, or with the presence of a transfer operon, they can be transferred readily among bacteria.

18-15. How do we determine if a bacterial phenotype is determined by a chromosomal or a plasmid gene?

Most directly, plasmids can be seen in the electron microscope and plasmid DNA can be isolated as a separate DNA fraction, distinct from the DNA of the host. Indirectly, the existence of plasmids can be demonstrated by the fact that many aspects of the phenotype tend to change simultaneously upon infection, and in fact, these aspects of the phenotype are infectious, another sure sign of the existence of a plasmid. There are also frequent spontaneous losses of plasmids due to their independent DNA replication. Loss would show up as a simultaneous loss of several aspects of the phenotype. Additionally, certain treatments are specific to plasmid DNA. For example, acridine dyes selectively prevent plasmid DNA replication thus resulting in loss of the plasmids from the cell population. Thus there are several direct and indirect methods to determine that plasmids are in a bacterial stock and responsible for aspects of the phenotype.

Key Concepts and Terms

Demes
Assortative mating
Inbreeding
Random genetic drift
Discrete generations
Linkage disequilibrium
Lethal-equivalent alleles
Autozygosity
Inbreeding coefficient, F
Path diagram

Populations
Random mating
Disassortative mating
Outbreeding
Gene pool
Linkage equilibrium
Common ancestry
Allozygosity
Identity by descent

Study Questions

19-1. The diagram below is of an electrophoretic gel stained for the autosomally controlled transferrin protein in mice. Bands are marked as slow (S) or fast (F). What are the frequencies of the two alleles?

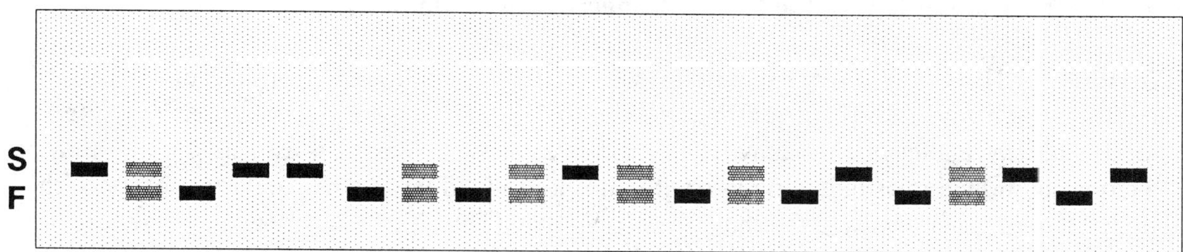

Each lane contains one of three protein patterns that is also indicative of the genotype. Thus when only a slow band is present, the genotype is *SS*; a fast band is indicative of the *FF* homozygote; and both bands is indicative of a *SF* heterozygote. This relationship has been demonstrated by breeding studies in which parents with slow bands produce offspring with slow bands; parents with fast bands produce offspring with fast bands; and parents with both types of bands produce offspring in ratios predicted on the basis of one locus with two alleles. By counting we get: Seven slow homozygotes (14 slow alleles), seven fast homozygotes (14 fast alleles), and six heterozygotes (six slow, six fast alleles). Thus by counting, we get 20 slow alleles out of 40 alleles, or an allele frequency of $p = f(S) = 0.5$, and $q = f(F) = 1 - p = 0.5$. We also could calculate first genotypic frequencies as follows: $f(SS) = 7/20 = 0.35$; $f(SF) = 6/20 = 0.30$; $f(FF) = 7/20 = 0.35$. Then $p = f(FF) + (1/2)f(FS) = 0.35 + 0.15 = 0.50$.

19-2. Given a population of human beings in which only the *A* and *B* alleles of the ABO blood system exist (no *O* allele is present). Of 300 people sampled in a population in Wyoming, 200 were type A (*AA* genotype), 75 were type AB (*AB* heterozygotes), and 25 were of type B (*BB* homozygote). What are the allelic frequencies?

Again we can count alleles. If there are 200 *AA* types, then they contribute 400 *A* alleles; similarly, type AB individuals contribute 75 each of the *A* and *B* alleles and type B individuals contribute 50 *B* alleles. Thus, $p = f(A) = (400 + 75)/600 = 0.7917$; $q = f(B) = 1 - 0.7917 = 0.2083$. We also could calculate first genotypic frequencies: $f(AA) = 200/300 = 0.667$; $f(AB) = 75/300 = 0.25$; $f(BB) = 25/300 = 0.083$. The frequency of the *A* allele is thus the frequency of type A homozygotes plus half the frequency of heterozygotes, or $p = f(A) = 0.667 + 0.125 = 0.792$, the same answer we got by counting alleles.

19-3. A geneticist is told by her assistant that in a population of 200 persons from Mobile, Alabama, 14% were type M (*MM* genotype), 57% were type MN (*MN* genotype), and 29% were type N (*NN* genotype). What are the frequencies of the *M* and *N* alleles?

In this case, the genotypic frequencies have already been calculated for you. You can use them directly or reconstitute the genotypic numbers by taking the appropriate percentages of 200 (14% of 200 is 28; thus 28 individuals were type M). However, given the genotypic frequencies, we calculate $p = f(M)$ as the frequency of *MM* homozygotes plus half the frequency of heterozygotes, or $0.14 + 0.285 = 0.425$. The frequency of the *N* allele, q, is $1 - p = 1 - 0.425 = 0.575$.

19-4. What are the assumptions of the Hardy-Weinberg equilibrium?

In order for Hardy-Weinberg equilibrium to hold, a group of assumptions about the lack of forces acting to perturb this equilibrium must hold. To begin with, we assume an autosomal locus with two alleles in a sexually reproducing population. Mating is assumed to be random. That is, an individual is likely to mate with another individual of a particular genotype with the probability of encountering that other individual. Thus if AA homozygotes make up 32% of a population, the probability of mating with an individual of that genotype is 0.32. Next, the population must be infinitely large in order to avoid sampling errors (random genetic drift) that occur in finite populations. Third, there should be no addition of alleles by migration nor change of alleles by mutation. And finally, no genotype should leave more offspring than any other genotype (no selection). Only after these conditions are met will a population be exactly in equilibrium.

19-5. What two methods can be used to prove that Hardy-Weinberg proportions are established after one generation of random mating?

In an infinitely large, sexually reproducing population, the frequencies of the three genotypes at an autosomal locus with two alleles will be distributed according to the binomial distribution ($p^2, 2pq, q^2$) if all of the other assumptions of the Hardy-Weinberg equilibrium are met. This can be shown using a demonstration or the gene pool concept. Using the gene-pool concept, we construct a Punnett square and add in probabilities. Thus, given that gametes from each sex are drawn at random to form gametes, each gamete has a probability of containing a particular allele equal to the frequency of that allele in the population. From this we see that genotypes are created in the proportion of p^2, $2pq$, and q^2 for the dominant homozygote, heterozygote, and recessive homozygote, respectively. A demonstration can also be used to show Hardy-Weinberg proportions by taking a population of individuals not in equilibrium and then simulating one generation of random mating. For example, if we assume that the three genotypes in a population occur in the frequencies X, Y, and Z (e.g., 0.1, 0.3, and 0.6, respectively) for *AA, Aa,* and *aa* genotypes, respectively, then we can construct a table of all possible matings, their frequencies (based on random mating), and the frequencies of the offspring. For example, the mating of *AA* males with *aa* females will occur as XZ (0.1 x 0.6 = 0.06). Since all offspring will be heterozygotes, XZ (6%) of the next generation will be heterozygotes from male *AA* by female aa matings. The frequencies of the offspring can be summed by genotype, the values of which can be factored. Thus there are X^2 + XY + (1/4)Y2 of *AA* offspring. This factors to (X +

[1/2]Y)2. X + (1/2)Y is in fact p^2; remember that an allelic frequency can be calculated as the frequency of homozygotes of that allele (X in this case) plus half the frequency of heterozygotes ([1/2]Y in this case). Similarly, the proportion of heterozygotes factors to 2pq and that for aa homozygotes to q^2, thus demonstrating that Hardy-Weinberg proportions are established by random mating regardless of the initial frequencies of the genotypes. You can show this by substituting numbers in the table for X, Y, and Z. The fact holds even if one value is zero.

19-6. Using the De Finetti diagram, show a population in Hardy-Weinberg equilibrium in which $p = 0.6$.

In the De Finetti diagram, a population is positioned by that unique point in which the perpendiculars to each side are in the proportions of the three genotypes. A parabola graphs those populations in Hardy-Weinberg equilibrium. Given that $p = f(A)$ and $q = f(a)$, you can thus see the point in the diagram in which $p^2 = 0.36$, $2pq = 0.48$, and $q^2 = 0.16$, given an altitude of unity. Since that population is in Hardy-Weinberg equilibrium, it falls on the parabola of equilibrium populations.

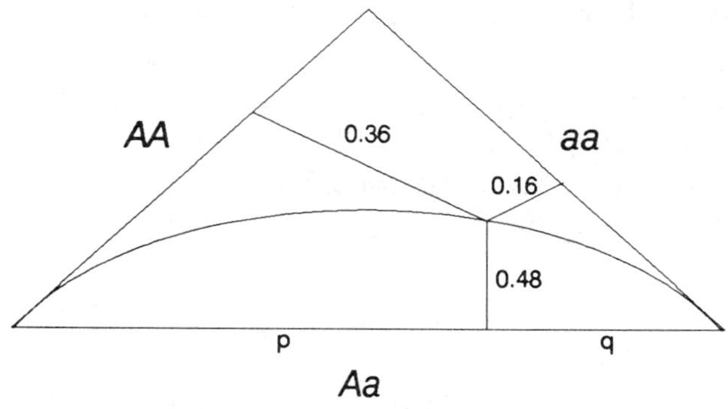

19-7. Is the population shown in Problem 19-1 in Hardy-Weinberg equilibrium?

When we ask this question, we are really asking whether the genotypes are inHardy-Weinberg proportions, p^2, $2pq$, and q^2. To determine this, we compare observed and expected numbers using the chi-square statistic. Remember that there were seven slow homozygotes, seven fast homozygotes, and six heterozygotes, and $p = f(S) = 0.5$ and $q = f(F) = 0.5$. We thus set up the following table:

```
                        SS      SF      FF      Total
Observed numbers        7       6       7       20
Expected proportions    p2      2pq     q2      1.0
                        (0.25)  (0.50)  (0.25)  1.0
Expected numbers        5       10      5       20
Chi-square value (O - E)2/E  0.8   1.6   0.8    3.2
```

Our expected proportions, given Hardy-Weinberg equilibrium, are p^2, $2pq$, and q^2, or 0.25, 0.50, and 0.25, for *SS, SF*, and *FF* genotypes, respectively. However, these are proportions of a total of 20 and thus each must be multiplied by the total (20) to get the expected numbers. We then subtract the observed (O) value from the expected value (E) for each category, square the value and then divide by E. These chi-square contributions are then summed for a total of 3.2, which is the chi-square value. That number must be compared with the critical value, the value at probability of 0.05 with one degree of freedom. Degrees of freedom are calculated based on the total number of categories, minus one for independent categories and minus one for each value estimated from the data. A rule of thumb, with population genetic data is that the degrees of freedom equal the number of phenotypes minus the number of alleles, in this case, 3 - 2 = 1. The critical value in the chi-square table is 3.841. Since our value (3.2) is less than this critical value we fail to reject our hypothesis (Hardy-Weinberg proportions). Since all expected values are equal to or greater than five, we can do the chi-square test.

19-8. Is the population shown in Problem 19-2 in Hardy-Weinberg equilibrium?

Remember that when we ask this question, we are really asking whether the genotypes are in Hardy-Weinberg proportions, p^2, $2pq$, and q^2. To determine this, we compare observed and expected numbers using the chi-square statistic. There were 200 *AA* homozygotes, 25 *BB* homozygotes, and 75 heterozygotes; $p = f(A) = 0.792$ and $q = f(B) = 0.208$. We thus set up the following table:

```
                        AA       AB       BB       Total
Observed numbers        200      75       25       300
Expected proportions    p2       2pq      q2       1.0
                        (0.627)  (0.329)  (0.043)  1.0
Expected numbers        188.1    98.7     12.9     300
Chi-square value        0.753    5.691    11.350   17.794
```

Our expected, Hardy-Weinberg proportions, are p^2, $2pq$, and q^2, or 0.627, 0.329, and 0.043, for *AA, AB,* and *BB* genotypes, respectively. However, these are proportions of a total of 300 and thus each must be multiplied by the total (300) to get the expected numbers. We then subtract the observed (O) value from the expected value (E) for each category, square the value and then divide by E. These chi-square contributions are then summed for a total of 17.794, the chi-square value. That number must be compared with the critical value, the value at probability of 0.05 with one degree of freedom. Degrees of freedom are calculated based on the total number of categories, minus one for independent categories and minus one for each value estimated from the data. The rule of thumb, with population genetic data, is that the degrees of freedom equal the number of phenotypes minus the number of alleles, in this case, 3 - 2 = 1. The critical value in the chi-square table is 3.841. Since our value (17.794) is greater than this critical value we reject our hypothesis (Hardy-Weinberg proportions). Since all expected values are equal to or greater than five, we can do the chi-square test.

19-9. Is the population shown in Problem 19-3 in Hardy-Weinberg equilibrium?

Remember that when we ask this question, we are really asking whether the genotypes are in Hardy-Weinberg proportions, p^2, $2pq$, and q^2. To determine this, we compare observed and expected numbers using the chi-square statistic. In this case we have to convert percentages into numbers. The 14% *MM*, 57% *MN*, and 29% *NN* out of 200 translates into 28 *MM*, 114 *MN*, and 58 *NN* individuals; $p = f(M) = 0.425$ and $q = f(N) = 0.575$. We thus set up the following table:

	MM	MN	NN	Total
Observed numbers	28	114	58	200
Expected proportions	p^2	$2pq$	q^2	1.0
	(0.181)	(0.489)	(0.331)	1.0
Expected numbers	36.2	97.8	66.2	200
Chi-square value	1.857	2.683	1.016	5.556

Our expected, Hardy-Weinberg proportions, are p^2, $2pq$, and q^2, or 0.181, 0.489, and 0.331, for *MM, MN,* and *NN* genotypes, respectively. However, these are proportions of a total of 200 and thus each must be multiplied by the total (200) to get the expected numbers. We then subtract the observed (O) value from the expected value (E) for each category, square the value and then divide by E. These chi-square contributions are then summed for a total of 5.556, the chi-square value. That number must be compared with the critical value, the value at probability of 0.05 with one degree of freedom. Degrees of freedom are calculated based on the total number of categories, minus one for independent categories and minus one for each value estimated from the data. The rule of thumb, with population genetic data, is that the degrees of freedom equal the number of phenotypes minus the number of alleles, in this case, 3 - 2 = 1. The critical value in the chi-square table is 3.841. Since our value (5.556) is greater than this critical value we reject our hypothesis (Hardy-Weinberg proportions). Since all expected values are equal to or greater than five, we can do the chi-square test.

19-10. In a sample of 100 persons from Chicago, 9 had blue eyes and the remaining 81 had brown eyes. Is the population from which this sample was drawn in Hardy-Weinberg proportions?

In this case, we assume that blue eyes are controlled by a recessive allele, b and the alternative, brown eyes, are controlled by the dominant allele, B. Since we cannot determine the proportion of heterozygotes among the brown-eyed individuals, we cannot count alleles to determine allelic frequencies. In fact, two phenotypes minus two alleles means zero degrees of freedom; we cannot do a chi-square test with zero degrees of freedom. However, if we assume that the population is in Hardy-Weinberg equilibrium, we can estimate allelic frequencies. If $p = f(B)$ and $q = f(b)$, the proportion of blue-eyed individuals (bb genotype) is q^2. Thus if we take the square root of the frequency of blue-eyed individuals we are estimating q. The square root of 0.09 is 0.3, which is our estimate of q, based on Hardy-Weinberg equilibrium. If $q = 0.3$, then $p = 0.7$ by subtraction. Therefore the proportion of homozygous brown-eyed individuals is $p^2 = (0.7)^2 = 0.49$, or 49 out of 100; the proportion of heterozygotes is $2pq = 2(0.7)(0.3) = 0.42$, or 42 of 100 persons. We cannot, however, verify this.

19-11. In a sample of 100 persons from Los Angeles, the following blood types were found: Type A, 17 (0.17); type B, 15 (0.15); type AB, 4 (0.04); type O, 64 (0.64). What are the allelic frequencies and is the population from which this sample was drawn in Hardy-Weinberg equilibrium?

Since dominance is involved, we cannot get allelic frequencies directly by counting--remember that type A individuals have AA and AO genotypes and type B individuals have BB and BO genotypes. However, there are four phenotypes and only three alleles (A, B, and O), giving one degree of freedom by subtraction, indicating that we can do a chi-square test. (We couldn't if there were zero degrees of freedom.) We will thus assume Hardy-Weinberg equilibrium and estimate allelic frequencies accordingly and then use the chi-square test to see if Hardy-Weinberg proportions hold. Although this sounds circular, you will see that there is at least one phenotypic class not needed in the calculation of allelic frequencies and thus, in a sense, independent. We let $p = f(A)$, $q = f(B)$, and $r = f(O)$. Thus, type O individuals should occur as r^2; therefore, the square root of the proportion of type O individuals gives us r. Since there were 64 of 100, the square root of 0.64 is 0.8, which is r. The sum of type A and type O individuals are three genotypes, AA, AO, and OO. Their frequencies combined are $p^2 + 2pr + r^2$, or $(p + r)^2$. Thus $(p + r)^2 = 0.17 + 0.64$ (types A + O) $= 0.81$. If we take the square root of both sides, $p + r = 0.9$. We already know that $r = 0.8$; if we subtract that from both sides, $p = 0.1$. Since $r = 0.8$ and $p = 0.1$, q must also be 0.1 since they all must add up to 1.0. We can now construct our chi-square table.

	Type A	Type B	Type AB	Type O	Total
Observed numbers	17	15	4	64	100
Expected proportions	p2 + 2pr	q2 + 2qr	2pq	r2	1.0
	(0.01 + 0.16)	(0.01 + 0.16)	(0.02)	(0.64)	1.0
Expected numbers	17	17	2	64	100
Chi-square value	0.0	0.235	2.0	0.0	2.235

In this case, we should not do the chi-square test because one of the expected values is less than five (type AB). One way that statisticians avoid this problem is to combine categories to eliminate an expected value less than five. We can thus redo the chi-square test as follows:

	Type A	Type B	Type AB + Type O	Total
Observed numbers	17	15	68	100
expected proportions	p2 + 2pr	q2 + 2qr	2pq + r2	1.0
	(0.01 + 0.16)	(0.01 + 0.16)	(0.02 + 0.64)	1.0
Expected numbers	17	17	66	100
Chi-square value	0.0	0.235	0.061	0.296

As you can see now, the chi-square value is smaller than the critical value (3.841 at one degree of freedom); thus the null hypothesis, Hardy-Weinberg proportions, is not refuted.

19-12. By test-crossing a dihybrid fruitfly (A and a alleles at the A locus and B and b alleles at the B locus) it is possible to determine the arrangement of alleles in the gamete of the dihybrid. That is, in the cross $AaBb$ x $aabb$, an offspring of the Ab phenotype must have been the result of an Ab gamete from the dihybrid. In this way, 100 gametic arrangements were scored by taking 100 females from a population cage of flies in which both alleles at both loci were present, and scoring one offspring from each female. It was found that there were: 66 AB arrangements (0.66); 4 Ab arrangements (0.04); 24 aB arrangements (0.24); and 6 ab arrangements (0.06). Is this population in linkage equilibrium?

Let $p_A = f(A)$, $q_A = f(a)$, $p_B = f(B)$, and $q_B = f(b)$. By counting alleles, we see that $p_A = 66 + 4 = 70/100 = 0.7$ and therefore $q_A = 0.3$; $p_B = 66 + 24 = 90/100 = 0.9$ and therefore, $q_B = 0.1$. If linkage equilibrium existed, we expect the AB arrangement to occur p_A x p_B of the time, or, 0.7 x 0.9 = 0.63 or 63 of 100. Similarly, we can construct a table of all observed and expected values:

	AB	Ab	aB	ab	Total
Observed	66	4	24	6	100
Expected	pApB	pAqB	qApB	qAqB	1.0
	(0.7)(0.9)	(0.7)(0.1)	(0.3)(0.9)	(0.3)(0.1)	1.0
	0.63	0.07	0.27	0.03	1.0
	63	7	27	3	100

As you can see the observed and expected numbers are not identical, but they are similar. The discrepancy can be quantified and treated statistically, but we will not do that here.

19-13. Given that second cousins mate and produce offspring. What is the inbreeding coefficient of such an offspring? Draw the pedigree and the path diagram of that mating.

The pedigree is converted into a path diagram by removing all extraneous individuals that cannot contribute alleles to both sides of the pedigree (family tree). In this case, there are two common ancestors, the common great grandparents of the second cousins. We

thus have two paths, one each for each common ancestor. In each path there are seven ancestors and no common ancestor is known to be inbred. Thus the inbreeding coefficient of the offspring of second cousins is: $F = 2(1/2)7 = 1/64 = 0.016$. In other words, the probability of autozygosity (identity by descent) is 1.6% at any locus, or 1.6% of the loci of the inbred individual are autozygous.

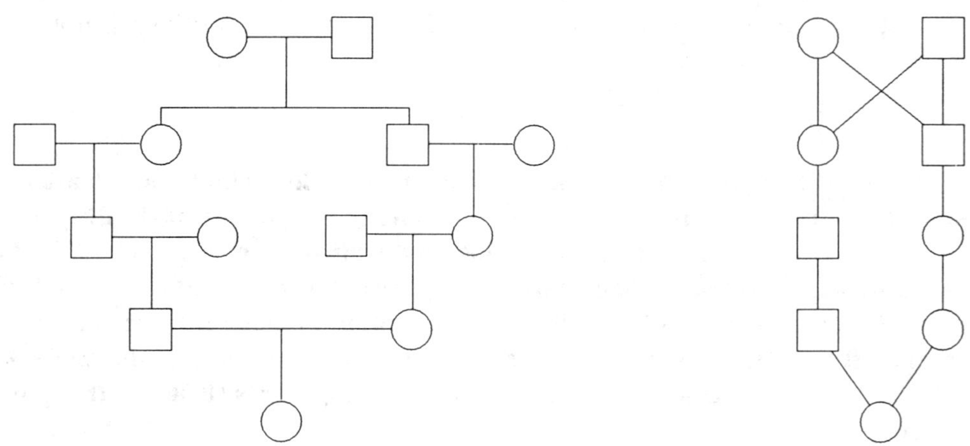

19-14. Two sisters marry two brothers and produce a son and a daughter. These children are double first cousins.

a) What is the inbreeding coefficient of one of their offspring?

b) Draw the pedigree and path diagram.

c) What would the inbreeding coefficient of the offspring of double first cousins be if the original sisters had inbreeding coefficients of 0.15?

Although more complicated, the rules of determining the inbreeding coefficient, F, are the same. In the pedigree, there are four common ancestors, each of the brothers and sisters in the original generation. Thus there are four paths, each with five ancestors.

a) The inbreeding coefficient of the offspring of double first cousins is: $F = 4(1/2)5 = 1/8 = 0.125$. Thus about 12.5% of the loci of the offspring are autozygous, or, there is about a 12.5% chance that any locus will be autozygous.

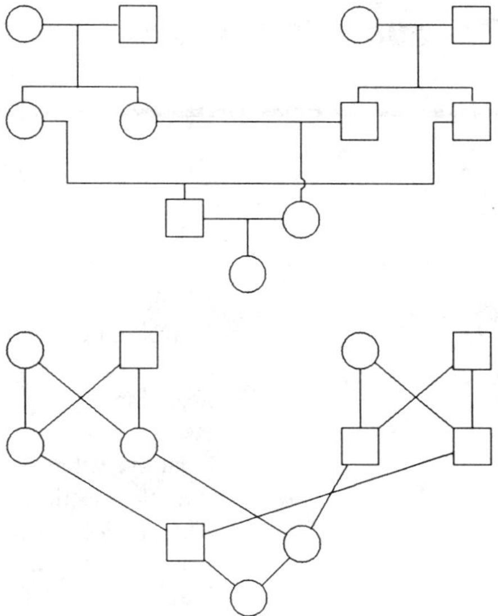

c) If we consider that the sisters had an inbreeding coefficient of 0.15 each, then the formula becomes: $F = 2(1/2)5(1.15) + 2(1/2)5 = 0.134$, an increase of 0.9%.

19-15. The MN blood types of a sample of 100 people is taken in Seattle. These are: 21 MM, 38 MN, and 41 NN. What is the population inbreeding coefficient and what does it mean?

In this case, $p = f(M) = (2[21] + 38)/200 = 0.4$ and $q = f(N) = 0.6$. Remember that the population inbreeding coefficient is defined as: $F = (2pq - H)/2pq$, in which H is the observed proportion of heterozygotes. In our case, $2pq = 2(0.4)(0.6) = 0.48$; H is 38% or 0.38. Thus, $F = (0.48 - 0.38)/0.48 = 0.208$. Therefore, in this population, heterozygosity (the proportion of heterozygotes) has been reduced by almost 21% due, presumably, to inbreeding.

Chapter 20 - Population Genetics: Processes That Change Allelic Frequencies

Key Concepts and Terms

Panmictic	Wahlund effect
Fokker-Planck equation	Deterministic
Bottlenecks	Founder effect
Natural selection	Selection
Fitness	Reproductive success
Selection coefficient	Adaptive value
Gametic selection	Zygotic selection
Fecundity selection	Sexual selection
Directional selection	Component of fitness
Disruptive selection	Stabilizing selection
Selection-mutation equilibrium	Mean fitness of the population, W
Heterozygote advantage	

Study Questions

20-1. What are the general procedures used to solve population genetics problems that involve intergenerational changes?

Many population genetics problems begin with a population in Hardy-Weinberg equilibrium that is then perturbed by some process; we usually need to determine the consequences of the perturbation. The equilibrium conditions are then needed and a determination needs to be made of whether these conditions are stable. In general, we begin the problem by putting the process in algebraic terms. Then we see what the new allelic frequency is by calculating the frequency after the process of interest has operated. From the difference in the two generations (before and after the process has acted) we can calculate change (delta-q). The equilibrium condition occurs when the change from one generation to the next is zero (at delta-$q = 0$). Stability is usually most easily examined by graphical means; the delta-q equation is graphed against all values of q.

20-2. Derive the equilibrium allelic frequency for an infinitely large, randomly mating population in which only forward and reverse mutation act to perturb the Hardy-Weinberg equilibrium.

Our algebraic model is set up such that we define forward mutation rate as u (A to a) and back mutation rate as v (a to A). The new equilibrium allelic frequency (q_{n+1}) is the old allelic frequency (q_n) plus additions due to forward mutation (up_n) minus losses to back mutation (vq_n), or: $q_{n+1} = q_n + up_n - vq_n$

Change in allelic frequency between generations is the new allelic frequency minus the old, or: delta-$q = q_{n+1} - q_n = q_n + up_n - vq_n - q_n = up_n - vq_n$

The equilibrium condition occurs when delta-q is zero, or: $upn - vqn = 0$, or $upn = vqn$

Since $p + q = 1$, $p = 1 - q$, or: $u(1 - q_n) = vq_n$

This can be solved for q_n, which at equilibrium is referred to as q-hat: $q\text{-hat} = u/(u + v)$ and $p\text{-hat} = v/(u + v)$.

20-3. In a particular population of bacteria, the forward mutation rate of the diaminopimelic acid decarboxylase locus (lysine requiring) is 3.3 X 10^{-5} (from the normal to the auxotrophic condition). The reverse mutation rate is 1.7 X 10^{-8} (from the auxotrophic condition to the wild type).

a) What is the equilibrium allelic frequency?

b) Why do these mutation rates differ so much?

c) Are our formulas valid for haploid bacteria?

a) We Use our formula for the equilibrium allelic frequency, q-hat: $q\text{-hat} = u/(u + v)$. In this case, we denote the wild-type allele as + and the mutant allele, *lys*. If $q = f(lys)$ and $p = f(+)$, then mutation of + to *lys* is the forward rate and mutation of *lys* to + is the reverse rate. Hence: $q\text{-hat} = 3.3 \text{ X } 10^{-5}/(3.3 \text{ X } 10^{-5} + 1.7 \text{ X } 10^{-8}) = 0.9995$. In other words, 99.95% of alleles in the population will be *lys* if no other process is in operation.

b) The two mutation rates differ by about three orders of magnitude because many, perhaps most, visible mutations of the wild-type allele convert it to the nonfunctional *lys* allele. However, for the *lys* allele to revert, only a very few base changes will return it to the wild-type state, most likely a change of precisely the same base back to the original.

c) Our formulas are valid for any organism. If you look at the formulas and how they were derived, only alleles in the gene pool and mutation rates of those alleles were considered. Ploidy is not a part of the model and thus the model holds regardless of the ploidy of the organism under study.

20-4. Is mutation equilibrium stable?

We examine equilibrium conditions by graphing the delta-q equation, the equation of change in allelic frequency, as a function of allelic frequency, q. When the delta-q equation is graphed (delta-q is the Y axis and q the X axis), the result is a straight line with a negative slope. This line indicates that the equilibrium is stable; if perturbed, the population will return to this point. You can see from inspection that when q is above the equilibrium point, change will be negative; and when q is below the equilibrium point, change is positive. Thus any perturbation of the allelic frequency from the equilibrium value causes it to return to this value. The graphed line intersects the delta-$q = 0$ line at this stable equilibrium point and the end points of the line, at $q = 0$ and $q = 1$, are the absolute values of the mutation rates.

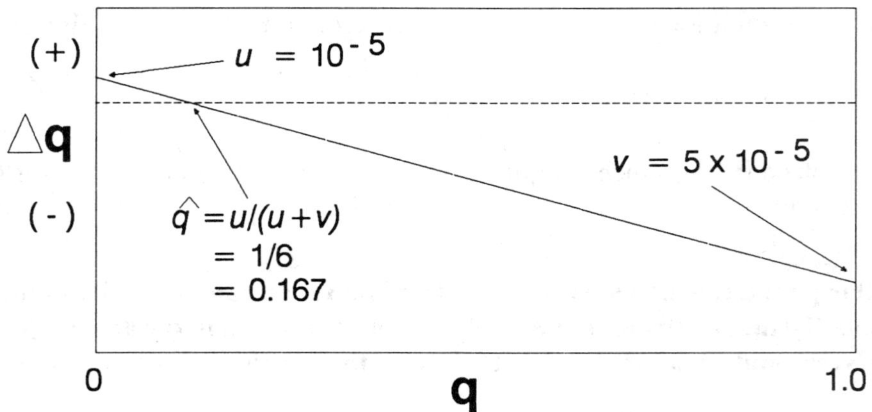

20-5. The population of natives on a small Pacific island had a frequency of the M allele of the MN blood system of 0.15. The population was then occupied during World War II by a population of soldiers whose allelic frequency was 0.60. After the occupying forces left, the next generation of islanders had an allelic frequency of 0.20. What was the migration rate of alleles into theslander's population?

The formula for allelic migration rate, m, is: $m = (q_c - q_1)/(q_2 - q_1)$ These values are allelic frequencies of: the conglomerate population, q_c; the original natives, q_1; and the migrants, q_2. Given our circumstances, the equation becomes: $m = (0.20 - 0.15)/(0.60 - 0.15) = 0.111$. This is interpreted to mean that 11.1% of the alleles in the Island population was contributed by individuals of the occupying force.

20-6. One thousand of the islanders from the previous problem were sampled for their MN blood type with the following results: type M, 90; type MN, 220; type N, 690. What factors might explain this distribution of genotypes?

With 90 MM genotypes, 220 MN genotypes, and 690 NN genotypes, we calculate allelic frequencies of $p = f(M) = 0.2$ and $q = f(N) = 0.8$. Based on Hardy-Weinberg proportions, we should have seen 40 M types, 320 heterozygotes, and 640 N types. If we do a chi-square test on this sample, we get a chi-square value of 97.656, which would cause us to reject our hypothesis of Hardy-Weinberg proportions. If we look carefully at the data we see there are too few heterozygotes and too many of each homozygote. A heterozygote deficiency can come about from several processes: inbreeding, selection against heterozygotes, and population subdivision (Wahlund effect). Given our knowledge of the MN blood system, we can probably rule out selection. Inbreeding would be a very likely explanation given our knowledge of what happens in small, isolated (island) populations. Also, given information about recent migration, there is the possibility that migrants and their new families formed a relatively isolated subpopulation on the island leading to subdivision effects. In order to know for certain what processes were in operation, more detailed studies would have to be undertaken.

20-7. Two populations of tropical fresh-water fish are observed. One is in a large lake with about a million individuals and the other is in a small pond with about 100 individuals. Both have a frequency of a malic dehydrogenase allele of 0.8. Assuming that no process other than sampling error is operating, what is the fate of the allele in each population?

Assuming that numbers will remain the same generation after generation, both populations have an 80% chance of ending up with a frequency of the allele of 1.0, and a 20% chance that the frequency will be 0.0. The dynamics of the process will be the same in both populations; the only thing different will be the time frame. In the small population, the changes will take place relatively rapidly whereas in the larger population, changes in allelic frequency caused only by sampling error (random genetic drift) will be slower. The ultimate fate is the same.

20-8. What processes can cause random genetic drift?

Random genetic drift is the change in allelic frequency that a population undergoes because the population is of finite size. The major cause is sampling error. That is, the gametes that make up a new generation are a sample of the gene pool of the parents. Since randomness is involved in the process of drawing these gametes during zygote formation, sampling error occurs. In other words, the allelic frequency of gametes will not usually be exactly the same as the allelic frequency of the parents. In addition, there will be random deaths of individuals that also can affect allelic frequency.

Several other processes lead to random genetic drift. Founder effects are changes in allelic frequency when a population is founded by a propagule (group of founders or migrants) that does not have the same allelic frequency as the parent population. Usually propagules are small, which is the major reason their allelic frequencies can vary considerably from the parent population. Bottlenecks of population density also lead to random genetic drift. When a population goes through one, with lowered density, sampling errors are exacerbated.

20-9. What are components of fitness?

Natural selection is seen as differential reproductive success; that is, some genotypes leave more offspring than others. However, the mechanism does not have to rest solely with superior reproduction, but can be affected by any aspect of the life cycle (component of fitness) of an organism on which natural selection can act. Generally, the action of natural selection on these components of fitness can be categorized as zygotic selection, gametic selection, sexual selection, and fecundity selection. Zygotic selection encompasses general survivorship; in fruit flies we talk of differential egg, larval, pupal, and adult mortality. Selection at any of these stages can alter allelic frequencies. Gametic selection is the differential success of one allele over another during gamete formation in a heterozygote. Sexual selection is mating success; some genotypes mate more than others. And, fecundity selection refers to fertility differences among different genotypes.

20-10. What are the outcomes of directional, stabilizing, and disruptive selection?

These three terms refer to the way in which natural selection can act on the phenotypic distribution of a population; it can favor one end of a distribution, the middle, or the ends. Directional selection is that which favors one end of a phenotypic distribution and results in a shift of that distribution over time. For example, natural selection seems to have favored larger brain size in hominid evolution. In stabilizing selection, natural selection acts to keep the distribution constant, with the same mean and variance. Thus it removes individuals at either end of the spectrum of the phenotype. In stable environments, most phenotypic traits are probably under stabilizing selection in which extremes are removed; most organisms are adapted to particular niches in which being too large or too small would be disadvantageous. Finally, disruptive selection acts against phenotypes in the middle of a distribution. For example, in African butterfly species that mimic more than one model species, selection favors phenotypes that resemble one model or the other, but not intermediate forms that do not mimic either model well. If we look at selection models, we can see that directional, stabilizing, and disruptive selection can be modeled as selection against the recessive phenotype, heterozygous advantage, and heterozygous disadvantage, respectively. This is a table of fitnesses:

```
                                          AA          Aa          aa
Selection against recessive homozygote     1           1         1 - s
Heterozygous advantage                   1 - s1        1         1 - s2
Heterozygous disadvantage                  1         1 - s         1
```

20-11. What is the ultimate fate of the a allele when heterozygotes are selected against?

Assuming that no other process is working on a particular population except heterozygous disadvantage, we can create the following selection model:

```
                        AA            Aa            aa          Total
Initial genotypic
   frequencies          p²           2pq           q²            1
Fitness (W)             1           1 - s          1
Ratio after selection   p²         2pq(1 - s)      q²         1 - 2pqs
                                                               (W-bar)

Genotypic frequencies
   after selection    p²/W-bar  2pq(1 - s)/W-bar  q²/W-bar      1
```

At any time in this process, the frequency of the a allele is q_n, and the frequency of the A allele is p_n. After one generation of natural selection, the new allelic frequency, q_{n+1}, is the frequency of the aa homozygote plus half the frequency of the heterozygote: $q_{n+1} = (q_n^2 + p_n q_n[1-s])/W$-bar

The difference in allelic frequency between generations, delta-q is: delta-$q = q_{n+1} - q_n = (q_n^2 + p_n q_n[1-s])/W$-bar $- q_n$

In order to subtract, we need to put q_n over the common denominator of W-bar by multiplying it by W-bar/W-bar:

delta-$q = (q_n^2 + p_n q_n[1-s])/W$-bar $- q_n W$-bar/W-bar

$\qquad = (q_n^2 + p_n q_n[1-s] - q_n W$-bar$)/W$-bar

We now substitute 1 - $2pqs$ for W-bar in the numerator:

$\qquad = (q_n^2 + p_n q_n[1-s] - q_n[1 - 2pqs])/W$-bar

The numerator is now expanded and simplified:

$\qquad = (q_n^2 + p_n q_n - s p_n q_n - q_n + 2pq2s)/W$-bar

$\qquad = (q_n^2 + p_n q_n - q_n - s p_n q_n + 2pq2s)/W$-bar

$q_n^2 + p_n q_n - q_n$ simplifies to zero; $- s p_n q_n + 2pq^2 s$ simplifies to $s p_n q_n(2q - 1)$. We thus simplify our delta-q equation to: delta-$q = s p_n q_n(2q - 1)/W$-bar. We then set this equation equal to zero to determine the equilibrium condition, q-hat, which occurs when any of the multipliers of the numerator are zero. Therefore, q-hat occurs at:

$\qquad$ s = 0

$\qquad$ p = 0

$\qquad$ q = 0

$\qquad$ 2q - 1 = 0

The first three would be considered trivial equilibria; that is there is no change because of either the absence of selection ($s = 0$) or the absence of one or the other of the alleles. The final equilibrium condition occurs at: $2q - 1 = 0$, or q-hat = 0.5 Thus there is an equilibrium at $p = q = 0.5$. Is this equilibrium stable? We can answer this question by graphing the delta-q equation for all values of q. In this case, the equilibrium is unstable. If q is above equilibrium, q will increase and if q is below equilibrium q will decrease. The reason why this is equilibrium is unstable is because natural selection is removing differentially heterozygotes; with each removal one of each allele is being removed. At $p = q$, each A and a allele is the same proportion of its group as the other is. However, at allelic frequencies higher or lower than equilibrium the rarer allele is a greater proportion of its alleles than the commoner allele is. Thus the removal of one of each allele is the equivalent of changing the allelic frequency to further lower the rarer allele.

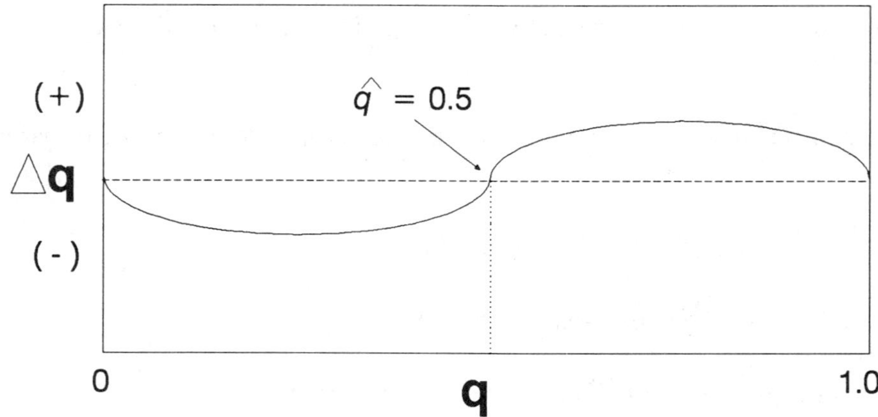

20-12. Given the following selection model, what is the equilibrium frequency of the *a* allele, *q*?

	AA	*Aa*	*aa*
Fitnesses	0.7	1	0.6

This is a model of heterozygous advantage; the heterozygote has the highest fitness (1). If that is the case, then algebraically, the fitness of the *AA* homozygote is $1 - s_1 = 0.7$; $s_1 = 0.3$. The fitness of the aa homozygote is $1 - s_2 = 0.6$; $s_2 = 0.4$. As you may remember, the equilibrium condition in heterozygous advantage is:

q-hat $= s_1/(s_1 + s_2) = 0.3/(0.3 + 0.4) = 0.43$
p-hat $= s_2/(s_1 + s_2) = 0.57$

20-13. In a given genetic system, the recessive homozygote is selected against with a selection coefficient, *s*, of 0.01. At the same time, mutation is bringing the *a* allele back into the population at a rate of $u = 5 \times 10^{-6}$. What is the equilibrium condition in this population if no other process is acting to perturb equilibrium?

This is an example of the model of selection-mutation equilibrium. The formula for the equilibrium condition is:

q-hat $\;=$ square root of u/s
$\phantom{q\text{-hat}\;}=$ square root of $5 \times 10^{-6}/0.01$
$\phantom{q\text{-hat}\;}=$ square root of 0.0005
$\phantom{q\text{-hat}\;}= 0.022.$

This is thus the equilibrium frequency of a deleterious allele being removed by natural selection but being restored by mutation. The frequency is low but at equilibrium

20-14. Why does it take so long to remove a deleterious allele from a population?

Assuming a situation in which the heterozygote is indistinguishable from the dominant homozygote, natural selection acts on deleterious recessive alleles only by removing recessive homozygotes. As allele frequency, $q = f(a)$, diminishes, the change per generation becomes less and less because recessive homozygotes are becoming rarer and rarer. However, the allele is becoming harder to remove because it is being found more and more in the heterozygous state as allelic frequency diminishes. At an allele frequency of $q = 0.5$, there are two heterozygotes for every recessive homozygote; at an allele frequency of $q = 0.1$, there are 18 heterozygotes for every recessive homozygote:

```
q            f(Aa) = 2pq      f(aa) = q²       f(Aa)/f(aa)
0.5            0.50             0.25               2
0.1            0.32             0.04               8
0.01           0.0198           0.0001            198
0.001          0.001998         0.000001        1,998
```

Hence the process of removing a deleterious allele, regardless of the magnitude of the selection coefficient, is asymptotic, taking an infinite number of generations to remove the last allele.

20-15. A sample of 100 fruit fly eggs was taken in a southwestern population in which a standard (ST) and inverted (AR) chromosome arrangement were segregating. When 100 adults were later sampled, the data in the following table resulted:

```
            ST/ST   ST/AR   AR/AR
Egg          30      50      20
Adult        26      60      14
```

What is the form of selection occurring here?

To understand the nature of selection, we need to calculate fitnesses. This is done by taking ratios of genotypes before and after selection has acted. These ratios are fitnesses; they can be standardized by dividing by the highest ratio to make the highest fitness equal to unity:

```
               ST/ST    ST/AR    AR/AR
Egg             30       50       20
Adult           26       60       14
Adult/Egg       0.87     1.2      0.7
Fitness (W)     0.73     1.0      0.58
s = 1 - W       0.27     0        0.42
```

From this table you can see that the fitnesses represent heterozygous advantage: the heterozygote has the highest fitness (standardized to unity). Selection coefficients are 0.27 against the ST/ST genotype and 0.42 against the AR/AR genotype. If $q = f(AR)$, $s_1 = 0.27$, and $s_2 = 0.42$, we can calculate the equilibrium condition:

q-hat $= s_1/(s_1 + s_2) = 0.27/(0.27 + 0.42) = 0.39$

In the adults, the allele frequency was 0.44, indicating that the population is not yet in equilibrium.

Chapter 21 - Genetics of the Evolutionary Process

Key Concepts and Terms

Survival of the fittest
Speciation
Phyletic evolution
Cladogenesis
Typological thinking
Reproductive isolating mechanisms
Hybrid zones
Sympatric speciation
Phyletic gradualism
Genetic polymorphism
Segregational load
Neutral gene hypothesis
Evolutionary rates
Phylogenetic tree
Group selection
Coefficient of relationship, r
Inclusive fitness
Haplodiploidy
Batesian mimicry
Industrial melanism
Cryptic coloration

Neo-Darwinism
Evolution
Anagenesis
Species
True speciation
Morphological species concept
Allopatric speciation
Parapatric speciation
Depauperate fauna
Punctuated equilibrium
Genetic load
Frequency-dependent selection
Clinal selection
Molecular evolutionary clock
Altruism
Sociobiology
Calculus of the genes
Kin selection
Mimicry
M.llerian mimicry

Study Questions

21-1. What was Darwin's concept of evolution by natural selection?

There are three main tenets to Darwin's model of evolution by natural selection:

> 1) variation exists

> 2) all populations overproduce young

> 3) the fittest will survive

The notion of variation being an integral part of a population was novel to Darwin; before that most naturalists saw each species as conforming to a "type" with variation being the exception. If every species produces more young than there is room for in the environment, then it stands to reason that those organisms that are better adapted to the environment have a better chance to survive and reproduce. Thus the variation inherent in a species can give some individuals an advantage over others in the struggle for existence. That is, the most fit will survive. To the extent that the variation is genetically controlled, the advantageous variation will be passed on to the next generation. In other words, the population evolves.

21-2. What are the differences between evolution and speciation?

Evolution is the change in allelic frequencies over time in a population; speciation occurs under two conditions of evolution. First, if a population branches into two populations that evolve to be different, speciation is said to have occurred when the two populations can no longer interbreed, a process called cladogenesis. Speciation can also occur when a species evolves over time until the current population is readily distinguishable from the progenitor population. This process is called anagenesis.

21-3. What are the differences between the biological and morphological species concepts? What are allopatric, parapatric, and sympatric speciation?

The biological species concept is based on the fact that two populations are called species if individuals from each cannot interbreed. This is a reasonable concept for diploid, outbreeding animals, but has many problems. For example, it cannot be applied to populations that are not in contact, or populations of fossil organisms, or populations in which the results of the interbreeding test are equivocal based on the fact that individuals that might not breed in the wild might breed in zoos or under laboratory conditions. For these reasons and more, the morphological species concept is frequently applied; we interpret species to be populations that are as different from each other morphologically as accepted species are different from each other. Thus we accept tigers and lions to be true species. If a new population were discovered that were as different from tigers and lions as tigers and lions are from each other, the new population would be called a new species based on the morphological species concept.

Allopatric, parapatric, and sympatric speciation are modes by which reproductive isolation (speciation) occurs. In allopatric speciation, a population is divided into two or more isolates by some geographical barrier. During isolation, the species evolve. In parapatric speciation, no obvious barrier arises, but part of the population at the edge of the species' range enters a new niche in which reproductive isolation evolves. Sympatric speciation occurs without any obvious barrier arising and occurs within the border of the species range. Possible examples are seen in host-specific insects that could switch hosts and become reproductively isolated on the new host.

21-4. What are the mechanisms of reproductive isolation?

Reproductive isolation must occur for two populations to become species. Many mechanisms can result in the inability of two formerly united groups to breed. These mechanisms are grouped frequently into prezygotic and postzygotic categories. Prezygotic mechanisms act to prevent gametes from uniting. These can be behavioral or mechanical differences in animals and spatial or temporal mechanisms in all organisms in which two populations occupy different habitats or are reproductively mature at different times of the year. Postzygotic mechanisms affect the hybrid organism after it has formed. These mechanisms include inviability or sterility of the hybrid.

21-5. Why are some cytogenetic change called "instantaneous speciation?"

Changes in the chromosomes of an organism that render it unable to successfully breed with its progenitor create instantaneous species. For example, assume a species with a chromosome number of 20 that undergoes somatic doubling to 40 chromosomes. Now instead of having gametes with 10 chromosomes, the gametes will have 20 chromosomes. Any fertilizations by the tetraploid of the parent species will produce triploid zygotes. If these are inviable or sterile, then the tetraploid is reproductively isolated. This mechanism is found more commonly in plants because they can exist vegetatively for an extended period of time and they usually don't have sex chromosomes, whose imbalance is often fatal to an organism.

21-6. What was the importance of Darwin's finches to the development of evolutionary theory?

Charles Darwin was the naturalist on board the ship, the H.M.S. Beagle, that charted coastal waters from 1831-1836, especially in South America, but also other parts of the world, including the Galapagos Islands. These islands, about 800 miles off the Ecuadorean coast in the Pacific Ocean, have been isolated for millions of years. The archipelago is divided into many volcanic islands, and thus the opportunity for evolution of the few species present was excellent. Especially noteworthy are the various species of birds that presumably evolved from a South American finch species and the many forms of the land tortoises present on the different islands. During his time on the islands, Darwin collected specimens of the birds and tortoises. He was struck by the similarities of the various species of birds (referred to now as Darwin's finches) on the islands. He wondered why mainland areas (e.g., South America) had so many different, unrelated bird species whereas the Galapagos Islands had species that were so similar to each other. Thinking along these lines, he was struck by the simple explanation that only one species colonized the archipelago and then it evolved into the many, yet similar, forms seen there today. Thus these finches had a great influence on the development of his theory of evolution.

21-7. What is punctuated equilibrium?

Darwin envisioned speciation (cladogenesis) as a slow, uniform process over time, with species gradually differentiating and becoming different from their sister or progenitor species (phyletic gradualism). In 1972, Niles Eldredge and Stephen Gould published a radically different view. They believed that cladogenesis takes place over a very short period of time with most of evolutionary time involved with little or no change in species. They called this view punctuated equilibrium: periods of stasis (equilibrium) punctuated by short bursts of rapid evolutionary change. These alternative views do not challenge the occurrence of evolution nor the way in which speciation occurs. Rather they question the tempo of speciation. These two views, phyletic gradualism and punctuated equilibrium, are relatively difficult to tell apart from the fossil record. Research is currently being done to determine the relative importance of each in evolutionary history.

21-8. How are genetic polymorphisms maintained in nature?

A genetic polymorphism is defined as the existence in the same population of two or more alleles of a locus. Usually some arbitrary value, say 1% or 5% is used as a limit: for there to be a polymorphism, the rarest allele must be at the cutoff percent or higher. Polymorphisms can be maintained by several mechanisms. First, they could be the result of heterozygous advantage in which two alleles are maintained. Second, they could be the result of transiency, in which a rare allele is in the process of becoming common through directional selection. Third, each allele could be favored in a different niche or at a different life-cycle stage resulting in the maintenance of both. Fourth, if the alleles are selectively neutral, they could be maintained by mutation followed by genetic drift. Fifth, the alleles could be maintained by frequency-dependent selection in which the fitness of an allele increases inversely with its frequency. Thus an allele is selected against when common but favored when rare. Some models of this kind of system have the alleles selectively neutral at intermediate frequencies, resulting in no selection at that time. Finally, there is the "other" category that includes any possible mechanism not yet thought of or mentioned. Currently, the consensus seems to be that most molecular polymorphisms (electrophoretic and DNA sequence variation) is selectively neutral.

21-9. What are hard and soft selection?

Hard and soft selection are two terms coined by Bruce Wallace. Hard selection is that in which selection coefficients are fixed values, unaltered by the environment. Tay-Sachs disease, for example, is caused by a lethal recessive allele. The homozygote has a selection coefficient of 1.0 (fitness of zero) regardless of other circumstances. However, soft selection refers to frequency- and density-dependent selection. In these two cases, selection coefficients are related to the frequency of an allele or the density of the population, respectively. In other words, a genotype that is not a good competitor in most ecological circumstances will have a low fitness. However, under very low density, without many competitors around, the individual with that genotype might have a very high fitness. This would be a case of density-dependent selection. An example of frequency-dependent selection is the case of rare-male advantage in fruit flies in which females seem to choose as males individuals with a rarer rather than a commoner genotype. When the favored male becomes common, females switch to preferring males of the now-rare genotype. Hence fitness depends on the frequency of an allele in the population. Cases that demonstrate both frequency- and density-dependent selection are referred to as cases of soft selection.

21-10. What is the neutral-gene hypothesis?

In the mid-1960s it was discovered with electrophoresis that there is a tremendous amount of variability (polymorphisms) in natural populations. Before that, it was assumed that polymorphisms were maintained by heterozygous advantage. However, for a polymorphism to be maintained that way, there has to be mortality or sterility of homozygotes. This loss might be tolerable to a population if there are only a few polymorphisms maintained that way. However, if there are thousands of polymorphisms in every individual, simple heterozygous advantage cannot be the mechanism of maintenance; there is too much of a loss of individuals to sustain a population. Thus scientists looked around for other explanations of all this variability. Motoo Kimura of Japan suggested that these polymorphisms might be selectively neutral. That is, electrophoretic variability might not be under selective control--all genotypes might do equally well. The variability could arise by mutation and be maintained by random genetic drift. This idea, at first rejected my many scientists, is now generally accepted as explaining much of the molecular variability in populations: electrophoretic and DNA sequence polymorphisms. This view does not deny that evolution and natural selection occur and that organisms are adapted to their environment. The view is simply that most variation is not selectively controlled.

21-11. What is a molecular evolutionary clock?

Many evolutionary biologists are concerned with determining the exact evolutionary history of life on earth. This includes very accurate phylogenetic trees, diagrams of the splitting of species from each other, and when these events took place. Before molecular techniques, taxonomists used morphological attributes of organisms to determine evolutionary relatedness and, primarily, fossil evidence to determine dates of splitting of species. Now, with molecular techniques of electrophoresis and protein and DNA sequencing it is possible to determine relatedness of organisms more precisely based on genetic similarity. Times of splitting can be determined directly if molecular changes among species occur at a uniform rate. If that is the case, then there is a molecular clock that can be used to determine precisely when species split from each other. In other words, if amino acid substitutions occur at a uniform rate, we merely count up the differences between two species in a given protein, divide by two because both species were evolving independently since the split, and multiply by the clock rate. However, a big unknown is whether the clock is uniform over time and over all aspects of the genome. Current evidence suggests that there is a uniform rate but that certain parts of enzymes may evolve more slowly and certainly there will be variability in the process. Neutralists have viewed the existence of a molecular clock as evidence of the neutral theory--a uniform clock must mean that most changes are not selective but simply time-dependent random processes.

21-12. Several proteins were examined in human beings and our closest living relatives, chimpanzees. Of 1,000 amino acids examined, we differed from the chimps by 23. What is the average number of amino acid substitutions per site, K? If the split took place 2.5 million years ago, what is the per-year rate, k?

The index K, the average number of amino acid substitutions, is defined as:
$$K = -\ln(1 - p)$$
where $p = d/n$ (d = number of amino acid differences and n = total number of sites looked at). In our example, $d = 23$, $n = 1,000$, and therefore, $p = 23/1,000 = 0.023$. Therefore, $-\ln(1 - 0.023) = -\ln(0.977) = 0.023$. Thus the average probability of substitution between human beings and chimps is about 2 percent per amino acid site. The per-year rate, k, is determined by dividing K by 2T, which in this case is 5 million years; therefore, $k = 4 \times 10^{-9}$. Thus that value is the average, per-year, per-site substitution rate between human beings and chimps. If there were a known molecular evolutionary clock rate, we could determine K and simply apply that clock rate to determine when human beings and chimps split rather than relying on fossil evidence for the 2.5 million year value.

21-13. What information do DNA sequences add to the neutralism controversy?

The proliferation of DNA-sequence information has added to two aspects of the neutralism controversy, both supportive of the neutralism view. First, DNA sequences have shown us that there is even more genetic variability than previously believed based on protein electrophoresis. With DNA sequences in front of us, it appears that every individual may be heterozygous at almost every locus. Certainly this level of variation cannot be maintained by selection. Second, given the nature of the genetic code, it is possible to make predictions based on the general level of variability in the different positions of the codons. That is, we expect fewer restrictions on the third position of the codon because of wobble rules. That is, there are nucleotide changes that can occur at the third position of codons that will not result in changes in amino acids in the protein and thus will have no phenotypic consequences. Empirical data support this prediction as well as the prediction that there will be more variability in introns that are never expressed and in other parts of the genome that do not get expressed directly. Hence DNA sequence data support the neutralism view.

21-14. How can natural selection maintain alleles for altruistic behavior?

Certain apparently altruistic behaviors, such as alarm calls in ground squirrels and caste systems in hymenoptera (failure to breed by workers in colonies and hives), could not readily be explained by natural selection acting on the fitness of individuals. This difficulty led to the suggestion of the model of group selection in which traits could be favored that were advantageous to the group (population or species). However, this view seemed inconsistent with the notion of individual selection; how could a trait be favored if it were disadvantageous to its bearer? This conundrum was solved by W. D. Hamilton who developed the notion of kin selection and inclusive fitness. Hamilton reasoned that if relatives shared genes, then natural selection could favor the genes for certain apparently altruistic behaviors if those behaviors led to an increase in those alleles in the next generation. For example, a ground squirrel that sounds an alarm call when a predator appears is more at risk from that predator than other squirrels in the population. However, if that alarm call were to warn close relatives of the caller, it could be selected for. In the case of caste systems in insects, almost all the cases known occur in hymenoptera that have a haplodiploidy sex-determining mechanism in which sisters are more closely related than a mother is to her own offspring. In that case, individual selection could favor those that helped their mother raise more sisters rather than raising their own offspring.

21-15. What are the differences between Batesian and Mullerian mimicry?

In both forms of mimicry, a mimic gains an advantage by looking like another organism, the model. In the case of Batesian mimics, innocuous organisms gain an advantage by fooling predators into thinking that they are dangerous or distasteful. For example, the viceroy butterfly is a mimic of the monarch butterfly. The monarch sequesters cardiac glycosides that it gets in its diet from milkweed that it eats. These chemicals make the monarchs highly noxious to predatory birds. Viceroys, on the other hand, do not sequester these compounds and are quite good to eat. By looking like the monarch the viceroy deceives predators that have encountered monarchs into avoiding the viceroys, thinking that the viceroys are monarchs. This system requires a noxious model to be in the vicinity of the mimic and for the models to be abundant. Otherwise, having too many mimics might teach the predators to try again after encountering a model.

Mullerian mimics on the other hand are groups of mimics, all of whom are noxious or dangerous, that gain an advantage by all looking alike. Thus they reinforce to potential predators that they are in a sense all the same species and to be avoided. An example is the general similarity of wasps and hornets. In this system, there is no requirement of one species being more common than another because all are both mimics and models.